THERMAL INFRARED SENSORS

THERMAL INFRARED SENSORS

THEORY, OPTIMISATION AND PRACTICE

Helmut Budzier and Gerald Gerlach

Dresden University of Technology, Germany

Translated by: Dörte Müller, *powerwording.com*

A John Wiley and Sons, Ltd, Publication

Library of Congress Cataloguing-in-Publication Data

Budzier, Helmut.
  [Thermische Infrarotsensoren. German]
  Thermal infrared sensors : theory, optimisation and practice / Helmut Budzier, Gerald Gerlach;
translated by Dörte Müller.
      p. cm.
  Includes bibliographical references and index.
  ISBN 978-0-470-87192-8 (cloth)
1.   Infrared detectors.   I. Gerlach, Gerald.   II. Title.
  TA1570.B83 2010
  621.36'2–dc22

                                                                        2010035043

A catalogue record for this book is available from the British Library.

Print ISBN: 9780470871928
E-PDF ISBN: 9780470976906
O-book ISBN: 9780470976913
E-Pub ISBN: 9780470976753

Set in 10/12 Times Roman by Thomson Digital, Noida, India

For Prof. Dr. Ludwig Walther, founder of the Dresden Infrared School

Contents

Preface

Until only a few decades ago, infrared technology was mainly the domain of military technology. In recent times, though, it has invaded an increasing number of new applications in our everyday lives. Examples are motion and fire detectors, ear thermometers, sensors that register the degree of browning in toasters, hand pyrometers for the contactless measuring of temperatures and thermal imaging devices. Infrared sensors are even the basis for new areas of application such as technical diagnosis, non-destructive evalution methods, environmental monitoring, gas sensors and remote sensing.

The technical interest in infrared radiation is due to the fact that it can be used both to determine the temperature without contact and thus the presence of bodies as well as the characteristics of bodies themselves including their structures:

- At room temperature, the maximum specific spectral radiation of blackbodies amounts to an approximate wavelength of 10 µm. This radiation wavelength range is therefore of fundamental importance for detecting real objects and determining their characteristics.
- The bond between the atoms of organic and anorganic molecules show resonance frequencies that almost always correspond to wavelengths in the infrared spectral range. If we can determine the frequency – or wavelength-related reflecting, transmitting and absorbing characteristics of substances and mixtures of substances – we can also determine the atomic or molecular structure of materials.

The increasing technical utilisation of infrared radiation in the mentioned areas of application is also related to central development trends in infrared measuring technology:

- Improved characteristics of infrared detectors. Research focuses particularly on increasing detectivity and improving the temperature resolution of such sensors as well as the transition to uncooled sensor principles.
- Development of highly integrated sensor arrays. Large pixel numbers of detector arrays require the miniaturisation of components and thus also the transition to semiconductor technology and the integration of sensor element and evaluation electronics. Thin layers on silicon substrates, the use of standard circuitry for evaluation electronics and the development of improved circuit technologies are of particular importance.
- Optimisation of infrared measuring systems. Here, the research focus is on the improvement of all system components and the optimisation of the characteristics of the total system.

- Analysis and development of new applications: contactless, emissivity-independent temperature measurements, spectroscopic applications, miniaturised spectroscopy, multicolour sensors, recognition systems and many more.

Thermal infrared sensors, in particular, are very important for civil applications as they can be used – as opposed to quantum detectors – in a non-cooled state and are therefore suitable for small and cost-efficient solutions and thus for large quantities.

Today, we do not only have a vast number of applications of thermal infrared sensors, but also the technological requirements regarding size, design, optical conditions, thermal and spatial resolution and many other framework conditions have diversified. This has resulted in very complex issues that users have to solve when trying to design optimum measuring arrangements or conditions. Each individual part of the measuring chain affects the relation between the source of the infrared radiation and the output signal of the measuring system. For this reason there are no simple rules and the problems cannot be solved without a basic understanding of the correlations.

There is a very limited number of textbook-like presentations available that summarise these issues. The present book intends to fill this gap by providing explanations of the essential basics of thermal infrared sensors and the correlation between the diverse effects. Using a large number of examples we will systematically show how this basic knowledge can be applied to the solution of specific tasks. Although the authors start with introducing the physical basics, they will only develop them to the point where they are necessary for real-life, recalculable ups with specific characteristics.

The goal of this book is to create a basic manual for users. It is intended to provide engineers, technicians, technical management, purchasers, and equipment suppliers with practical knowledge regarding the usage of modern infrared sensors and measuring systems. The book focuses on thermal infrared sensors. This way, we want to avoid exceeding the scope and keep it handy. The authors agreed that this would not constitute any serious restriction. On the one hand, it is mainly the uncooled, thermal sensors that represent the largest increases in commercial sales and determine the major part of new applications. On the other hand, it is possible to transfer major parts of the information to quantum detectors.

The present book is based on lectures on infrared measuring technology that the authors have held during many years at the Technische Universität Dresden, Germany. This book, however, was designed in a completely different way in order to turn the information into a basic user manual. This means this publication has been a new experience for us, too. We are aware that it will not be complete right away, and would therefore appreciate corrections, ideas, and suggestions for improvements (Helmut.Budzier@tu-dresden.de; Gerald.Gerlach@tu-dresden.de).

The authors would like to express their thankfulness to Volker Krause, Ilonka Pfahl and the publisher, in particular Simone Taylor and Nicky Skinner who provided the necessary support for a fast and uncomplicated publishing process.

We would like to express our appreciation to Dörte Müller who did a great job in translating the manuscript from German into English.

We would also like to thank Volkmar Norkus (TU Dresden/Germany), Norbert Neumann (InfraTec GmbH Dresden/Germany), Günter Hofmann (DIAS Infrared GmbH Dresden/Germany), Jörg Schieferdecker (Heimann Sensor GmbH Dresden/Germany) and Jean-Luc Tissot (ULIS France) for their support, discussions and for providing materials.

Dresden, June 2009
Helmut Budzier, Gerald Gerlach

List of Examples

List of Symbols

a	dimension; acceleration; pulse amplitude; thermal diffusivity
b	dimension
c	heat capacity
d	distance; diameter
d_P	piezoelectric coefficient
e	unit vector; elementary charge
e	natural number
f	frequency; function; focal length
g	acceleration of gravity
h	height; PLANCK constant
i	numeration index; electric current
$\tilde{i}$	effective or rms value of the noise current
j	imaginary unit; numeration index
k	factor; coefficient; direction component; wave number; wave vector; numeration index; discrete state
k_B	BOLTZMANN constant
l	length; direction component
m	dimension of an array; coefficient; mass; direction component; numeration index
n	number; refractive index; dimension of an array; charge carrier density; direction component; numeration index
p	pressure; hole density; probability
q	charge; heat flux density
r	number; curvature radius; position vector
s	modulus of elasticity; complex frequency; layer thickness; standard uncertainty; path
t	duration; time
u	numeration index
v	velocity
$\tilde{v}$	effective or rms value of the noise voltage
w	probability density
x	coordinate
y	coordinate
z	coordinate
z_{AB}	quality factor
A	area; cross-section

B	bandwidth; width; magnetic flux density
C	capacitance; constant
D	dielectric displacement; diameter; detectivity
E	irradiance; modulus of elasticity; electric field strength; energy; expectation value
E_a	activation energy
E_g	energy of the band gap
F	force; form factor; focal plane
FOM	Figure of Merit
H	irradiation; magnetic field strength; transfer function; principal plane
I	electric current; radiant intensity
J	current density
K	parameter
L	radiance; scene size
M	exitance mass; molar mass
N	number
N_A	AVOGADRO constant
P	power; polarisation; probability
Q	charge; radiation energy
R	distance; responsivity; universal gas constant; resistance; ratio
S	strain; power density; POINTING vector
T	stress; temperature; period
$\tilde{V}$	effective value of signal voltage
V	voltage; volume
W	energy; distribution function
X	driving variable; input variable
$\overline{X}$	average or mean value of X
$\underline{X}$	complex value of X
Y	output variable
Z	impedance
α	absorbance; coefficient of linear thermal expansion; angle; parameter
α_B	temperature coefficient of resistance
α_S	SEEBECK coefficient
β	parameter; angle; complex wave number
χ	susceptibility
δ	DIRAC delta function
γ	angle
ε	dielectric constant; emissivity; permittivity
π	LUDOLPHine number (3.14159...)
η	efficiency
κ	conductivity
λ	wavelength; thermal conductivity; event rate;
θ	angle
μ	permeability; chemical potential
ν	frequency; pulse rate
π_P	pyroelectric coefficient
ρ	density; charge density; reflectance; resistivity

σ	STEFAN–BOLTZMANN-constant; distribution
τ	lifetime of charged particles; transmittance; time constant
φ	angle
ω	angular frequency; projected solid angle
Δ	deviation; change; difference; path difference
Φ	radiant flux; radiant power
ϑ	temperature in $^\circ$ C (absolute temperature T –273.15 K)
Π	product
Σ	sum
Ω	solid angle
Ω_0	solid angle unit (= sr)

Indices

a	ambient, acceleration
b	width
Ch	Chopper
d	thickness
eq	equivalent
f	frequency-specific
g	geometrical
i	inner
l	length; idle
m	mechanical; mean
max	maximum
min	minimum
n	electrone; normalised
on	on
off	off
p	hole
r	noise, random
rect	rectangular
rn	normalised noise
th	thermal
x	coordinate direction
y	coordinate direction
z	coordinate direction
A	absorption; numeration index
B	bolometer; band; numeration index
BB	blackbody
BLIP	background limited infrared performance
C	capacitator
CC	circular cone
E	electric; electric field strength; edge; environment
F	Fermi
HS	half-space

I	current
L	load; conduction band
O	object, observation
P	pyroelectric
Q	charge
R	radiation, random
S	sensor; surface; surroundings
Si	silicon
T	temperature
V	voltage; valence band; volume

| λ | wavelength-specific; spectral |
| v | frequency-specific |

| 0 | amplitude; output value; reference; resonance; ambient |
| 1, 2, 3 | coordinate direction; parameter |

Abbreviations

ac	alternating current
ACF	autocorrelation function
BB	blackbody
BLIP	background limited infrared performance
CCF	cross-correlation function
dc	direct current
ESF	knife-edge spread function
FOV	field of view
FPA	focal plane array
IFOV	instantaneous field of view: *FOV* of one pixel of an array
LSF	line spread function
MTF	modulation transfer function
NEP	noise-equivalent power
NETD	noise-equivalent temperature difference
OTF	optical transfer function
PSF	point spread function
PTF	phase modulation function
SiTF	signal transfer function
SNR	signal-to-noise ratio

1

Introduction

1.1 Infrared Radiation

1.1.1 Technical Applications

Infrared (IR) radiation is an electromagnetic radiation in the wavelength range between visible radiation (often abbreviated as VIS; $\lambda = 380$–780 nm) and microwave radiation ($\lambda = 1$ mm–1 m). IR radiation has some physical characteristics that make them particularly suitable for a number of technical applications:

- Each body emits electromagnetic radiation (see Section 2.3). The radiation depends on the wavelength and is determined by the body's temperature. Thus the measured radiation can be used to measure the body's temperature. This characteristic is used for contactless temperature measurement (pyrometry).
- For high temperatures of several 1000 K, the maximum falls within the visible range; the human eye has adapted its highest sensitivity to $\lambda \approx 550$ nm, corresponding to the surface temperature of the sun (approximately 6000 K). Opposed to this, at ambient temperature the irradiation of bodies has an infrared maximum of about 10 μm (see Figure 3.2.1). This can be used to detect the presence and motion of people (motion detectors, security systems) or to record entire scenes with IR cameras – similar to video cameras. The latter has the advantage that parts of the IR spectrum allow the propagation of radiation even in darkness or under foggy conditions – the basis for night vision devices and driver assist systems.
- IR cameras can also be used for recording thermal images showing the thermal isolation of buildings, temperature distribution of combustion processes or temperature-dependent processes. Today, commercial thermal cameras have an image resolution similar to that of high-resolution TV.
- Electromagnetic radiation can induce oscillations in the atoms of molecules. In that case distance and angle of the bonds between atoms, for instance, change periodically. Each bond has a specific resonance frequency at which the radiation is almost completely absorbed.

Thermal Infrared Sensors: Theory, Optimisation and Practice   Helmut Budzier and Gerald Gerlach
© 2011 John Wiley & Sons, Ltd

As radiation frequency v and wavelength λ are coupled via propagation velocity c.

$$\lambda = \frac{c}{v} \qquad\qquad (1.1.1)$$

Chemical compounds absorb radiation at characteristic wavelengths. Many of these absorption wavelengths fall within the IR range. IR radiation with a specific wavelength can thus be used to determine the presence and concentration of specific substances, which can be applied for the gas analysis. If we record complete reflexion and transmission spectres of irradiated samples, the position of the absorption bands can be used to draw conclusions about their chemical composition (IR spectrometry).

Looking at the mentioned characteristics and the corresponding technical applications, the typical structure of infrared measuring systems becomes apparent (Table 1.1.1). The

Table 1.1.1 Typical structure of infrared sensors and measuring systems

Application	Radiation source	Propagation path	Imaging system	Sensor/Sensor array
Pyrometer	Measuring object (with background)	Atmosphere, optical fibre	Lens, filter	Single- or multi-element sensor (mainly thermal or pyroelectric)
Thermal imaging device	Measuring object with background/ Thermal imaging scene	Atmosphere	Lens, filter	Sensor array (FPA Focal Plane Array, mainly bolometer)
Passive motion detector	Measuring object with background	Atmosphere	Fresnell-optics	Two-element sensor (mainly pyroelectric)
Gas sensor	Thermal (e.g. glow emitter, hotplate) or non-thermal emitter (LED, Laser)	Gas cell or atmosphere with measuring gas	Lens, filter	Multi-element sensors (mainly thermal or pyroelectric)
Spectrometer	Thermal (e.g. glow emitter) or tunable, non-thermal emitter (LED, Laser)	Chemical compound	Grating, mirror	Single-element or array sensors (mainly thermal or pyroelectric)
Relevant book chapters	Chapter 2	Section 2.1	Chapter 3, Section 5.5	Section 2.1, Chapter 4 to 6

measuring object can be the IR radiation source itself (pyrometry, thermal imaging, motion detectors) or affect the transmission of the propagation path (gas analysis, spectroscopy/ spectrometry).

The structure of this book follows the measuring chain presented in Table 1.1.1, which means that Chapter 2 will discuss the origin and propagation of electromagnetic radiation. The radiation sources will be limited to thermal emitters as they themselves constitute the measuring object in pyrometers, thermal imaging devices as well as motion detectors and are preferably used in gas analysis and spectrometry.

Section 2.1 discusses the effect that the propagation of electromagnetic radiation occurring on the propagation path has, particularly on the detection of chemical species.

Chapter 3 presents the photometric basics including mapping the radiation source area to the area of the sensor or sensor array. As in most applications the IR radiation is emitted from an emitter's surface into space; we will put particular emphasis on the solid angle relations between radiation source and sensor. Due to the huge variety, classical optical elements such as lenses, gratings or filters will not be included, as this is not a book on optics. An exception will be made in Section 5.5, which will introduce the optical parameters that are important for sensor arrays.

Chapter 5 describes the characteristics of infrared optical sensors and sensor arrays. As the minimum detectable radiant fluxes or temperature differences, respectively, are determined by physically unavoidable noise processes, Chapter 4 will introduce the basics and the most important noise sources for IR sensors.

Chapter 6 describes the structure and characteristics of important thermal infrared sensors. We limit the discussion to thermal IR sensors as they do not have to be cooled and therefore can be miniaturised and are comparatively inexpensive (thermal imaging cameras with HDTV resolution are currently already available for only several thousand to tens of thousands Euro). They have come to clearly dominate the civil applications' market.

Chapter 7 finally presents an overview of the basics presented in the previous chapters for the applications included in Table 1.1.1.

1.1.2   Classification of Infrared Radiation

Infrared radiation is a high-frequency electromagnetic radiation. For the propagation in linear-optical components (vacuum, air, glass, silicon), frequency v remains constant, whereas wavelength λ can change depending on wave propagation (light) velocity c in different media:

$$\lambda = \frac{c}{v} = \frac{c_0}{n\,v} \tag{1.1.2}$$

where c_0 is the speed of light in vacuum and n the refractive index. In spectroscopy, wave number σ is often used as the reciprocal of the wavelength:

$$\sigma = \frac{1}{\lambda} \tag{1.1.3}$$

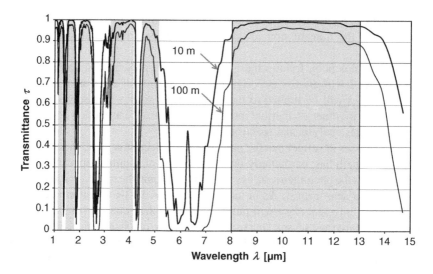

Figure 1.1.1 Typical transmission in atmosphere during summer in Central Europe. Parameter: Length of propagation path

Wavelength ranges can be classified according to several criteria. In the following, we will present the classification that is commonly used in infrared measuring technology and that results from transmission τ of the atmosphere due to the absorption of water vapour (H_2O) and carbon dioxide (CO_2) in the air (Figure 1.1.1). For all applications where the atmosphere constitutes the transmission path, only selected wavelength ranges – atmospheric windows – can be used (areas shaded in grey in Figure 1.1.1; Table 1.1.2).

Table 1.1.2 Atmospheric windows

Atmospheric window	Infrared region	Wavelength range in µm	Temperature range of max. exitance acc. to Equation. 2.3.6 in K	Remarks
I	Near infrared (NIR)	1.2–1.3	2415–2229	Emission of hot bodies (>1000 °C)
		1.5–1.7	1932–1705	
		2.1–2.4	1380–1208	
II	Mid infrared (MIR)	3.2–4.1	906–707	Emission of hot bodies (>300 °C)
		4.4–5.2	659–557	Absorption of CO_2 and other gasses
III	Far infrared (FIR)	8–13	362–223	Emission of bodies at room temperature

Table 1.1.3 Classification of infrared radiation

Range		Wavelength λ (μm)	Wave number σ (cm^{-1})	Frequency $(\nu)^a$ (THz)	Photon energy E^b (eV)
VIS		0.38–0.78	26 316–12 821	789–384 THz	3.27–1.59
IR	NIR	0.78–3	12 821–3333	384–100 THz	1.59–0.41
	MIR	3–6	3333–1667	100–50 THz	0.41–0.21
	FIR	6–40	1667–250	50–7.5 THz	0.21–0.03
	UFIR	40–1000	250–10	7.5 THz–300 GHz	0.03–1.2×10^{-3}

$$^a\nu/\mathrm{Hz} = \frac{299.8 \times 10^{12}}{\lambda/\mu\mathrm{m}}$$

$$^bE/\mathrm{eV} = \frac{1.241}{\lambda/\mu\mathrm{m}}$$

In the optimal case, the maximum of the radiation source's specific irradiation lies exactly in the selected range (WIEN'S displacement law; Equation. 2.3.6).

In correspondence to the atmospheric windows in Table 1.1.2, the infrared radiation range can be divided into near- (NIR), mid- (MIR), far- (FIR) and ultrafar- (UFIR) infrared (Table 1.1.3).

Here, monochromatic radiation is radiation of a single frequency or wavelength, respectively. Mostly though, radiation consists of many wavelengths (or frequencies) and we have to look at the corresponding spectral ranges.

1.2 Historical Development

Table 1.2.1 summarises the historical development of infrared measuring technology. The starting point was the discovery by Sir WILHELM HERSCHEL in 1800, that for the spectral decomposition of light the largest temperature rise occurred in the invisible spectral range beyond the red (*Philosophical Transactions of the Royal Society of London*, **XIII**, April 24, 1800, page 272, quoted in Caniou [1]). Later on he showed that also invisible radiation from other hot sources such as fire, candlelight or a red-hot oven emit invisible radiation that behaves according to the laws of optics regarding reflexion and diffraction. Initially this radiation was called 'ultra-red', but later the term 'infrared' was introduced.

Further development stages focused on the proof that thermal radiation and electromagnetic waves are of the same nature (first half of the nineteenth century). MAX PLANCK'S works formulating the light-quantum hypothesis and the derivation of PLANCK'S radiation laws in 1900 as well as the formulation of the law of the external photoeffect in 1905 and the assumption of stimulated emission by ALBERT EINSTEIN constituted the decisive fundamentals for the quantum nature of the interaction of electromagnetic radiation and solid-state bodies. Thus, the essential physical bases for the technical utilisation of infrared radiation in IR measuring technology were established.

Table 1.2.1   Milestones in the development of infrared measuring technology (according to [1] and [2])

Year	Event
1800	Discovery of the existence of thermal radiation in the invisible beyond the red by W. HERSCHEL
1822	Discovery of the thermoelectric effects using an antimony–copper pair by T. J. SEEBECK
1830	Thermal element for thermal radiation measurement by L. NOBILI
1833	Thermopile consisting of 10 in-line Sb-Bi thermal pairs by L. NOBILI and M. MELLONI
1834	Discovery of the PELTIER effect on a current-fed pair of two different conductors by J. C. PELTIER
1835	Formulation of the hypothesis that light and electromagnetic radiation are of the same nature by A. M. AMPÈRE
1839	Solar absorption spectrum of the atmosphere and the role of water vapour by M. MELLONI
1840	Discovery of the three atmospheric windows by J. HERSCHEL (son of W. HERSCHEL)
1857	Harmonisation of the three thermoelectric effects (SEEBECK, PELTIER, THOMSON) by W. THOMSON (Lord KELVIN)
1859	Relationship between absorption and emission by G. KIRCHHOFF
1864	Theory of electromagnetic radiation by J. C. MAXWELL
1879	Empirical relationship between radiation intensity and temperature of a blackbody by J. STEFAN
1880	Study of absorption characteristics of the atmosphere through a Pt Bolometer resistance by S. P. LANGLEY
1883	Study of transmission characteristics of IR-transparent materials by M. MELLONI
1884	Thermodynamic derivation of the STEFAN law by L. BOLTZMANN
1894, 1900	Derivation of the wavelength relation of blackbody radiation by J. W. RAYEIGH and W. WIEN
1903	Temperature measurements of stars and planets using IR radiometry and spectrometry by W. W. COBLENTZ
1914	Application of bolometers for the remote exploration of people and aircrafts
1930	IR direction finders based on PbS quantum detectors in the wavelength range 1.5–3.0 μm for military applications (GUDDEN, GÖRLICH and KUTSCHER), increased range in World War II to 30 km for ships and 7 km for tanks (3–5 μm)
1934	First IR image converter
1939	Development of the first IR display unit in the US (Sniperscope, Snooperscope)
1947	Pneumatically acting, high-detectivity radiation detector by M.J.E. GOLAY
1954	First imaging cameras based on thermopiles (exposure time of 20 min per image) and on bolometers (4 min)
1955	Mass production start of IR seeker heads for IR guided rockets in the US (PbS and PbTe detectors, later Sb detectors for Sidewinder rockets)
1965	Mass production start of IR cameras for civil applications in Sweden (single-element sensors with optomechanical scanner: AGA Thermografiesystem 660)
1968	Production start of IR sensor arrays (monolithic Si-arrays: R.A. SOREF 1968; IR-CCD: 1970; SCHOTTKY diode arrays: F.D. SHEPHERD and A.C. YANG 1973; IR-CMOS: 1980; SPRITE: T. ELIOTT 1981)
1995	Production start of IR cameras with uncooled FPAs (Focal Plane Arrays; microbolometer-based and pyroelectric)

1.3   Advantages of Infrared Measuring Technology

Infrared radiation has a number of advantages that make it very useful for contactless temperature measuring, in particular, and infrared measuring technology, in general.

- It is contactless and therefore (almost) reactionless. As the radiation energy exchange also takes place from the sensor to the radiation source, it would not be correct to talk about a complete absence of feedback. However, this feedback on the radiation source can – in general – be neglected (see Section 6.1 and Figure 6.1.3).
- It spatially separates radiation source and detector, which means that also very hot or otherwise difficult-to-access objects can be measured.
- It allows fast measurements as it propagates at the speed of light and the characteristic time constants of measuring process can be kept very small, due to miniaturisation (see Example 6.2.2 and Figure 6.2.6, for instance).
- It allows measuring of the temperature of a solid-state body surface and not that of the surrounding atmosphere.

These characteristics enable the contactless examination or temperature measurement of the following objects and measuring points:

- Fast-moving objects. Contactless measuring avoids interferences caused by contacting temperature sensors, rotation or friction.
- Current-carrying objects. Current-carrying components and devices pose a potential risk for both measuring equipment and operating staff. Contactless remote sensing can be used to avoid such risks.
- Small objects. The larger the temperature sensor is in relation to the object to be measured the more the temperature measured by the temperature sensor increases. Due to the adjustable imaging system (see Table 1.1.1) of IR measuring system, such interference can be largely avoided.
- Measuring at unaccessible measuring points. Many industrial measuring processes are carried out under very harsh conditions, for example at high temperatures that would destroy contacting temperature sensors. The IR-based contactless measuring technology provides technical solutions for such measuring tasks.
- Parallel measurements at several measuring points. In case of contact sensors, measuring processes with measurements at several points require complex solutions. Sensing measuring systems or image-based measuring procedures are a much more (cost-)efficient solution for such measuring tasks (quantitative thermograms or thermal imaging).

1.4   Comparison of Thermal and Photonic Infrared Sensors

In principle, we distinguish two kinds of infrared sensors (Figure 1.4.1):

- thermal sensors and
- photon or quantum sensors.

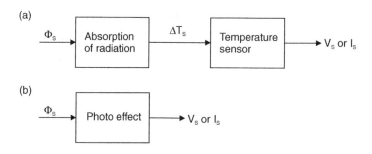

Figure 1.4.1 Active principle of (a) thermal and (b) photon or quantum radiation detectors, respectively

Thermal sensors are radiation detectors that, due to the absorption of IR radiation, experience a change in temperatures and convert it into an electric output signal (Figure 1.4.1a). They are therefore often called radiation temperature sensors. Looking at a sensor with current output I_S, without loss of generality, the following relation results for the current responsivity (index I):

$$R_I = \frac{\Delta I_S}{\Delta \Phi_S} \qquad (1.4.1)$$

with change ΔI_S of the current sensors due to the change of radiant flux $\Delta \Phi_S$ (unit Watts) absorbed in the sensor (see Equation 5.1.2). For energetic reasons, a specific radiant flux causes a proportional temperature increase ΔT_S in the sensor and subsequently, a proportional output current ΔI_S. Theoretically, energy absorption and thus temperature dependency ΔT_S is independent of wavelength, which means that also R_I is unrelated to wavelength. In practice, radiation absorption shows a certain wavelength dependency, though.

Specific detectivity D^* shows the relation resulting from responsivity R_I and the effective value of noise currents $\tilde{i}_{Rn}$ that falsify the sensor output current; and it is therefore a measure of the signal-to-noise ratio (SNR). D^* is normalised to the root of sensor surface A_S:

$$D^* = \frac{\sqrt{A_S}}{\tilde{i}_{Rn}} R_i \qquad (1.4.2)$$

Apart from the dynamics of the incident IR radiation, the result is a wavelength-independent effective value of the sensor output current and thus a wavelength-independent, specific detectivity D^* of thermal sensors:

$$D^* \neq D^*(\lambda) \qquad (1.4.3)$$

It is only natural to observe that the wavelength-dependency of the absorption of sensor materials causes a wavelength-dependency of responsivity R_I or specific detectivity D^*, respectively, of thermal sensors.

As opposed to this, radiant flux Φ_S impinging on photon detectors causes a completely different wavelength-dependency.

The incident radiant flux (radiation power) Φ_S on the sensor corresponds to radiation energy dQ impinging on the sensor per time unit dt (for a detailed presentation, see Section 2.2):

$$\Delta\Phi_S = \frac{dQ}{dt} \tag{1.4.4}$$

Radiation energy Q is the sum of energy contributions $h \cdot v$ of N photons:

$$\Delta\Phi_S = \frac{dQ}{dt} = \frac{d(N \cdot h \cdot v)}{dt} = hv\frac{dN}{dt} \tag{1.4.5}$$

For quantum detectors, each photon on average contributes η electrons to the current conduction. η is the quantum efficiency. Sensor current ΔI_S thus becomes

$$\Delta I_S = \frac{d(\text{Charge})}{dt} = \frac{d(\eta Ne)}{dt} = \eta e\frac{dN}{dt} \tag{1.4.6}$$

It results from Equations 1.4.5 and 1.4.6 that current responsivity R_I of a quantum detector is (see Section 5.1):

$$R_I = \frac{\Delta I_S}{\Delta\Phi_S} = \frac{\eta e}{hv} = \frac{\eta e}{hc}\lambda \tag{1.4.7}$$

Thus, the responsivity of photon sensors increases with increasing wavelength. This can be illustrated if we look at the radiation of two different wavelengths λ_1 and $\lambda_2 = 2\lambda_1$:

$$Q = N_1 hv = N_1\frac{hc}{\lambda_1} = N_2\frac{hc}{\lambda_2} \tag{1.4.8}$$

If radiation energy Q is the same in both cases, radiation of wavelength λ_2 contains exactly twice as many photons

$$N_2 = N_1\frac{\lambda_2}{\lambda_1} = 2N_1 \tag{1.4.9}$$

because its photon energy $h \cdot v_2$ is only half as large as $h \cdot v_1$:

$$h \cdot v_2 = h\frac{c}{\lambda_2} = h\frac{c}{2\lambda_1} = \frac{1}{2}(h \cdot v_1) \tag{1.4.10}$$

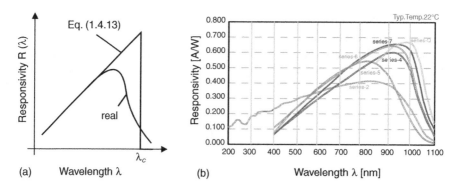

Figure 1.4.2 Wavelength-dependent responsivity of photon sensors. (a) schematic representation, (b) Si-pin-photo diodes (by courtesy of Silicon Sensor GmbH), the serial numbers characterise photo diodes with specifically optimised characteristics

Problems arise if wavelength λ is very large and thus the photon energy too small for the energy to be sufficient to transport the electrons from the valence band via energy gap E_G to the conduction band:

$$h \cdot v = \frac{hc}{\lambda} < E_G \qquad (1.4.11)$$

or respectively

$$\lambda > \frac{hc}{E_G} \qquad (1.4.12)$$

In this case, there are no conduction electrons available any longer and the responsivity becomes zero:

$$R_I = \begin{cases} \dfrac{\eta e}{hc} \lambda & \text{for } \lambda \leq \lambda_C = \dfrac{hc}{E_G} \\[3mm] 0 & \text{for } \lambda > \lambda_C = \dfrac{hc}{E_G} \end{cases} \qquad (1.4.13)$$

where λ_C is the cut-off wavelength.

Figure 1.4.2a shows a schematic representation of the current responsivity curve of quantum sensors. In principle, this allows the following conclusions:

- The responsivity of photon sensors is, in principle, wavelength-dependent.
- Photon detectors have a restricted operating wavelength range that is mainly determined by the band structure or band gap E_G. For silicon with $E_G = 1.1 \, \text{eV}$, the cut-off wavelength is

$$\lambda_C = \frac{hc}{E_G} = \frac{6.626 \times 10^{34} \, \text{W s}^2 \cdot 2.998 \times 10^8 \, \text{m/s}}{1.12 \, \text{eV}} = 1.11 \, \mu\text{m}$$

This is the reason why silicon can only be used for detectors in the near-IR range, whereas the MIR- and FIR-range also can apply other semiconductors.

Table 1.4.1 summarises the differences of the characteristics of thermal and photon sensors.

Table 1.4.1 Comparison of central characteristics of thermal and photon sensors

Parameter	Thermal sensors	Photon sensors
Responsivity		
Detectivity D^*	• Wavelength-dependence only. determined by radiation absorption • Wavelength-independent. • Temperature-dependence as power of the temperature. • Only marginal improvement of D^* by cooling.	• Wavelength proportional to λ up to cut-off wavelength λ_C. • Heavily wavelength-dependent (Figure 1.4.1). • Very large D^* values are achievable in the NIR range. • Mostly exponential temperature dependency. • Cooling can largely increase D^*.
Operation	Uncooled	Cooled
Frequency range	Up to several 100 Hz	Up to GHz (photo diodes, photo cells)

Example 1.1: Responsivity and Detectivity of Thermal Sensors and Photon Sensors

The resolution limit of an infrared sensor is absolutely determined by the noise of the detected radiation, that is of the signal to be measured. For thermal sensors it is the radiation noise (Section 4.2.4) and for photon sensors the photon noise. The resulting maximum detectivity is called background-limited infrared photodetection (BLIP) and for thermal sensors background-limited infrared performance. BLIP detectivity for thermal sensors will be presented in Section 5.3. It amounts to:

$$D^*_{\text{BLIP,TH}} = \frac{1}{\sqrt{16\varepsilon k_B \sigma T_S^5}} \qquad (1.4.14)$$

with emissivity ε, BOLTZMANN constant k_B, STEFAN–BOLTZMANN constant σ as well as sensor temperature T_S. The result for $T_S = 300\,\text{K}$ is, according to Example 5.7, a wavelength-independent BLIP detectivity of $D^*_{\text{BLIP,TH}} = 1.81 \times 10^8\,\text{m} \cdot \text{Hz}^{\frac{1}{2}}\,\text{W}^{-1}$.
  For photo sensors, it applies that [1]:

$$D^*_{\text{BLIP,PH}} = \frac{\lambda}{hc}\sqrt{\frac{\eta}{2Q_B}} \qquad (1.4.15)$$

with quantum efficiency η, wavelength λ and integral photon exitance of background Q_B:

$$Q_B = \sin^2 \frac{FOV}{2} \int_0^{\lambda_C} Q_{\lambda s} \, d\lambda \qquad (1.4.16)$$

Here, FOV is the field of view (see Section 5.5, Figure 5.5.1). Analogously to spectral exitance $M_{\lambda s}$ of a blackbody thermal emitter (see Equation 2.3.3), it applies to the spectral photon exitance that

$$Q_{\lambda s} = \frac{c_1'}{\lambda^4} \frac{1}{e^{\frac{c_2}{\lambda T_B}} - 1} \qquad (1.4.17)$$

with radiation constants c_1' and c_2 and background temperature T_B.

In the near infrared range, the BLIP detectivity of photon sensors is – by several orders of magnitudes – larger than the detectivity of thermal sensors. In far infrared, that is in the range of maximum radiation of bodies in the ambient temperature range, both have approximately the same magnitude (Figure 1.4.3).

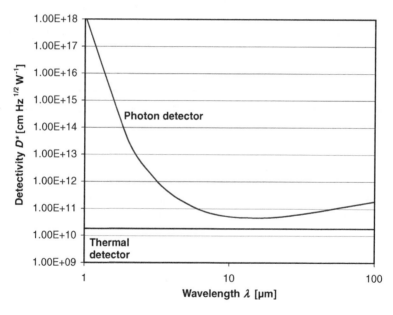

Figure 1.4.3 BLIP detectivity of thermal and photon sensors $FOV = 180°$; $\eta = \varepsilon = 1$; $T_B = T_S = 300$ K

1.5 Temperature and Spatial Resolution of Infrared Sensors

The structure of infrared sensors and measuring systems presented in Table 1.1.1 shows that the radiation energy of a measuring object or a radiation source – affected by the transmission characteristics of the propagation path – can be mapped onto a single-element sensor or an array sensor (FPA). The following practical questions are important:

1. What is the minimum temperature difference of radiation source or measuring object against the background of a single sensor element, and
2. What is the minimum spatial distance between two points with a certain temperature difference that can still be detected by the array sensor.

The temperature resolution is determined by the noise-equivalent temperature difference (*NETD*; see Section 5.4). This is the object's internal temperature difference for which, in the sensor, the noise signal is equivalent to the measuring signal (signal-to-noise ratio $SNR = 1$). *NETD* applies to each individual pixel (sensor element) of a sensor array. The smaller the solid angle that is projected by the imaging system to one sensor pixel, the smaller the sensor signal and thus also *NETD*. Smaller IR measuring systems with a smaller imaging system usually result in a lower-quality temperature resolution.

The modulation transfer function (*MTF*; see Section 5.6) describes the smallest presentable structure and is therefore a measure of the spatial resolution. In principle the spatial resolution is also affected by thermal and electric couplings between the pixels of a sensor array. Optical diffractions and the ratio between pixel dimension and structural size to be imagined have a particularly large influence, though. Especially regarding the latter it becomes obvious very soon that pixels are not able to distinguish the points of the source any longer if the pixel size is larger than half the spatial period of the points.

In principle, a better spatial resolution can be only achieved by smaller pixels. However, this results in a lower-quality temperature resolution, as described above.

Temperature and spatial resolution are two completely different characteristics of sensor arrays. An optimisation of both characteristics partially requires changes of design parameters that are opposed to each other, as the example of the pixel size shows.

1.6   Single-Element Sensors Versus Array Sensors

Section 5.1 to 5.5 present the derivation of sensor parmeters for individual sensor elements that can work both as a separate single-element sensor and an individual pixel of a matrix array sensor (focal plane array, FPA). Section 5.6 will then discuss – separately for array sensors – the spatial resolution of IR sensors, which is only relevant for multi-element sensors.

As Figure 1.6.1 shows that for the signalling chain of a system for contactless temperature measurement, it is theoretically possible to treat sensors with a single sensor surface (single-element sensors) and sensors with many individual sensor surfaces (multi-element sensors, array sensors) exactly in the same way regarding thermal resolution. Each pixel of a thermal imager measures the temperature of a small area of the object and constitutes itself a pyrometer. However, there are different boundary conditions for sensors in pyrometers and in thermal imaging devices. An essential difference is the available sensor surface (field of view). It is practically determined by the optics. For pyrometers, it may amount up to several square millimetres. For pyrometers, the field of view equals the sensor surface. Thermal imaging devices commonly have fields of view with a diagonal of up to 20 mm. In imaging systems, mainly optical distorsion and vignetting restrict the field of view. Only a small

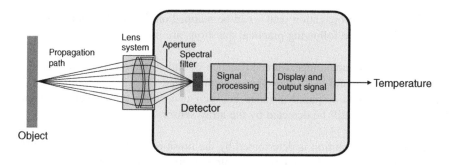

Figure 1.6.1   Signalling chain of a system for contactless temperature measuring

portion of it is available to a pixel of the sensor. With an increasing field of view, also the optical devices and sensors have to be designed larger – which results in a rather substantial increase of the system's total costs. For this reason, the sensor surface should be kept as small as possible. From the sensor's perspective it should be as large as possible, though (Equation 5.4.7).

All the presented types of thermal sensors can be manufactured both as single elements or as array sensors. Examples will, however, only include typical arrangements, such as pyroelectric dual sensors or microbolometer arrays. The micro-bridge technology, for instance, which is presented in the section on microbolometer arrays, is also available for pyroelectric and thermoelectric sensors.

References

1. Caniou, J. (1999) *Passive Infrared Detection, Theory and Application*, Kluwer Academic Publishers, Dordrecht.
2. Herrmann, K. and Walther, L. (1990) *Wissensspeicher Infrarottechnik (Store of Knowledge in Infrared Technology)*, Fachbuchverlag, Leipzig.

2

Radiometric Basics

Thermal sensors measure the energy or power (energy per time unit) of the absorbed part of the incident electromagnetic radiation. As this absorbed radiation power is converted into heat it causes the sensor to become warmer. Therefore, thermal sensors have at least one radiation absorber and one sensor element that converts the temperature increase into an electric output signal. In principle, thermal sensors thus constitute a special group of temperature sensors, radiation temperature sensors. Pyrometry – the contactless temperature measurement of a body – uses the body's own thermal radiation in the infrared spectral range of the electromagnetic radiation. Looking at the physical processes occurring at thermal sensors, we have to take into consideration the propagation of the radiation between measured body and sensor as well as the propagation and absorption of the radiation in the sensor before we can discuss the operational principle of thermal sensors.

2.1 Effect of Electromagnetic Radiation on Solid-State Bodies

2.1.1 Propagation of Radiation

MAXWELL'S equations and the material equations describe the propagation of radiation (Table 2.1.1):

MAXWELL'S equations:

$$\text{rot }\vec{H} = \frac{\partial \vec{D}}{\partial t} + \vec{J} \quad \text{Ampere's law} \tag{2.1.1}$$

$$\text{rot }\vec{E} = -\frac{\partial \vec{B}}{\partial t} \quad \text{Faraday's law} \tag{2.1.2}$$

$$\text{div }\vec{D} = \rho \quad \text{Gaussian law} \tag{2.1.3}$$

$$\text{div }\vec{B} = 0 \tag{2.1.4}$$

Thermal Infrared Sensors: Theory, Optimisation and Practice   Helmut Budzier and Gerald Gerlach
© 2011 John Wiley & Sons, Ltd

Table 2.1.1 Explanations of variables in MAXWELL'S and material Equations 2.1.1–2.1.9

Term	Symbol	Unit
Magnetic field strength	$\vec{H}$	A/m
Electric field strength	$\vec{E}$	V/m
Dielectric displacement	$\vec{D}$	$C/m^2 = As/m^2$
Polarisation	$\vec{P}$	$C/m^2 = As/m^2$
Magnetic flux density	$\vec{B}$	Vs/m^2
Current density	$\vec{J}$	A/m^2
Charge density	ρ	$C/m^3 = As/m^3$
Vacuum permittivity	ε_0	F/m = As/(V m)
Relative permittivity	ε_r	1
Conductivity	κ	A/(Vm)
Vacuum permeability	μ_0	H/m = Vs/(A m)
Relative permeability	μ_r	1
Electric susceptibility	χ	1
Mechanical stress	T_m	N/m^2
Piezoelectric coefficient	d_P	As/N
Pyroelectric coefficient	π_P	C/(K m) = As/(K m)

Material equations:

$$\vec{B} = \mu_0 \mu_r \vec{H} \qquad (2.1.5)$$

$$\vec{J} = \kappa \vec{E} \qquad (2.1.6)$$

$$\vec{D} = \varepsilon_0 \vec{E} + \vec{P} \qquad (2.1.7)$$

with

$$\vec{P} = \chi \varepsilon_0 \vec{E} + d_P T_m + \vec{\pi}_P \Delta T \qquad (2.1.8)$$

$$\chi = \varepsilon_r - 1 \qquad (2.1.9)$$

Three of the four MAXWELL'S equations (Equations 2.1.1–2.1.4) are unrelated; the fourth derives from the others. The material equations (Equations 2.1.5–2.1.7) are often falsely subsumed under MAXWELL'S equations, but they describe the relationship between different field quantities in different materials. These equations mostly constitute a rough estimation as there are often non-linear and partially even lossy and hysteretic relationships. For the propagation of light, the latter often can be neglected. Losses in materials, such as dissipation, are often described using complex instead of real material parameters.

For the high-frequency propagation of electromagnetic radiation in z-direction, there is a specific solution to the MAXWELL'S equations:

$$E_x(z, t) = E_0 \exp\{j(\omega t - kz)\} \exp\left\{-\frac{\alpha}{2}z\right\} \qquad (2.1.10)$$

with E_x being the electric field strength in z direction, k the wave number, and α the absorption.

For wavenumber k, it applies that:

$$k = \frac{\omega}{c} = \frac{\omega n}{c_0} = \frac{2\pi}{\lambda} \qquad (2.1.11)$$

Wavenumber k is the modulus of wave vector $\vec{k}$. The direction of wave vector $\vec{k}$ corresponds to the propagation direction of the electromagnetic wave. Light speed c in a medium is calculated from the speed of light *in vacuo* and refractive index n of the medium:

$$c = \frac{c_0}{n} \qquad (2.1.12)$$

The material parameters are used to calculate the speed of light:

$$c = \frac{1}{\sqrt{\mu\varepsilon}} \qquad (2.1.13)$$

Thus, it applies for the refractive index that:

$$n = c_0\sqrt{\mu\varepsilon} = \sqrt{\mu_r\varepsilon_r} \qquad (2.1.14)$$

This way, the optical characteristics of a medium are determined by their electric or dielectric characteristics, respectively, and vice versa. Using absorption α, we can define penetration depth z_α of an electromagnetic radiation into matter as the decrease to value 1/e:

$$z_\alpha = \frac{2}{\alpha} \qquad (2.1.15)$$

Penetration depth z_α for copper, for instance, is approximately 5 nm and in glass about 10^4 m. If there is no absorption (lossless media: $\alpha = 0$), it applies that:

$$E_x(z,t) = E_0 \exp\{j(\omega t - kz)\} \qquad (2.1.16)$$

Analogous definitions apply to the magnetic field strength. It results from MAXWELL'S equations (Equations 2.1.1–2.1.4) that the direction of the electric and magnetic field strength vector are perpendicular. For the same position, it is therefore possible to determine the magnetic component from the electric field strength:

$$H_{0y} = n\sqrt{\frac{\varepsilon_0}{\mu_0}}E_{0x} \qquad (2.1.17)$$

Thermal (and also all optical) sensors, however, cannot be directly determined from the field strengths. They measure the power transported in the field which results from the POYNTING vector $\vec{S}$ [1]:

$$\vec{S} = \vec{E}x\vec{H} \qquad (2.1.18)$$

As the frequencies of infrared light (wavelengths $\lambda = 780$ nm–1 mm correspond to a lightwave frequency of $v = c/\lambda = 385$ THz–300 GHz) are much larger than the characteristic frequencies of the sensor, the sensor is not able – due to the time constant – to follow the instantaneous value of the amplitude, but it produces an average over time. The temporal average value of the POYNTING vector corresponds to intensity I of the radiation:

$$I = \bar{S} = \left|\vec{E}x\vec{H}\right| \qquad (2.1.19)$$

As magnetic and electric field strengths are perpendicular, it applies that:[1]

$$I = \bar{S} = \overline{E(t)H(t)} = \lim_{T \to \infty} \frac{1}{T} \int_{t}^{t+T} E_0 H_0 \cos^2(\omega t)\,\mathrm{d}t \qquad (2.1.20)$$

The solution of the integral is:

$$I = \frac{1}{2} E_0 H_0 = \frac{1}{2} E_0^2 n \sqrt{\frac{\varepsilon_0}{\mu_0}} \qquad (2.1.21)$$

The intensity units are W/m^2 and the intensity can be determined with just one field component, electric or magnetic field strength. It is therefore sufficient to analyse only one of the two field components, for example the electric field strength.

2.1.2   Propagation in Lossy Media

As mentioned above, for lossy media with radiation absorption α it is possible to define a complex wave number (propagation constant) $\underline{\beta}$ instead of real wavenumber k:

$$\underline{\beta} = k - j\frac{\alpha}{2} \qquad (2.1.22)$$

Thus, Equation 2.1.16 becomes

$$E_x(z,t) = E_0 \exp\{j(\omega t - \underline{\beta} z)\} \qquad (2.1.23)$$

For plane waves in the z direction it results from MAXWELL'S equations (Equations 2.1.1 and 2.1.2) as well as from Equations 2.1.6 and 2.1.7 that

$$\frac{\partial^2 \vec{E}_x}{\partial z^2} = \mu\varepsilon \frac{\partial^2 \vec{E}_x}{\partial t^2} + \mu\kappa \frac{\partial \vec{E}_x}{\partial t} \qquad (2.1.24)$$

Transforming Equation 2.1.24 into the image area, the result is

$$(j\underline{\beta})^2 E_0 \exp(-j\underline{\beta} z) = \{\mu\varepsilon(j\omega)^2 + \mu\kappa j\omega\} E_0 \exp(-j\underline{\beta} z) \qquad (2.1.25)$$

or

$$\underline{\beta}^2 = \mu\varepsilon\omega^2 - \mu\kappa j\omega \qquad (2.1.26)$$

By equating the absolute values and phases on the left-hand and the right-hand side of the complex Equation 2.1.26 and the subsequent solution of a quadradic equation, it results for wavenumber k that

$$k^2 = \frac{\mu\varepsilon\omega^2}{2} \left\{ \sqrt{1 + \left(\frac{\kappa}{\omega\varepsilon}\right)^2} + 1 \right\} \qquad (2.1.27)$$

---

[1] This statement only applies to elementary fields. In anisotropic materials, for instance, angles can assume any value. In that case, Equation 2.1.19 has to be used to calculate the mean value via the cross product.

and for absorption α

$$\left(\frac{\alpha}{2}\right)^2 = \frac{\mu\varepsilon\omega^2}{2}\left\{\sqrt{1+\left(\frac{\kappa}{\omega\varepsilon}\right)^2}-1\right\} \tag{2.1.28}$$

If the medium has a small absorption α, it applies that

$$\kappa \ll \omega\varepsilon \tag{2.1.29}$$

and the expression under the root in Equation 2.1.28 becomes approximately 1 and $\alpha \approx 0$. The same applies, for instance, for dielectrics with $\kappa \approx 0$. It results for the wavenumber from Equation 2.1.27 that

$$k^2_{\text{Dielectric}} \approx \left(\frac{\omega}{c}\right)^2 \tag{2.1.30}$$

For good conductors, such as metals, it applies that $\kappa \gg \omega\varepsilon$ and therefore the wavenumber becomes:

$$k^2_{\text{Metal}} \approx \frac{\mu\kappa\omega}{2} \tag{2.1.31}$$

with increasing conductivity κ, propagation speed c and wavelength λ decrease.

In dielectrics, it usually applies that $\mu_r \approx 1$, and Equation 2.1.14 becomes:

$$n = \sqrt{\varepsilon_r} \tag{2.1.32}$$

Analogously to complex wavenumber $\underline{\beta}$, it is possible to define a complex refractive index for lossy media

$$\underline{n} = n - jn' \tag{2.1.33}$$

with extinction coefficient n' and a complex permittivity:

$$\underline{\varepsilon_r} = \varepsilon'_r - j\varepsilon''_r \tag{2.1.34}$$

By inserting Equations 2.1.33 and 2.1.34 in Equation 2.1.32 and squaring, we get

$$n^2 + n'^2 - j2nn' = \varepsilon'_r - j\varepsilon''_r \tag{2.1.35}$$

which can be used to calculate the real and imaginary parts of the permeability:

$$\varepsilon'_r = n^2 + n'^2 \tag{2.1.36}$$

$$\varepsilon''_r = 2nn' \tag{2.1.37}$$

Inserting Equation 2.1.34 into Equation 2.1.26, we arrive at:

$$\underline{\beta}^2 = \omega^2\mu\varepsilon' - j\omega\mu(\kappa + \omega\varepsilon'') = \omega^2\mu\varepsilon' - j\omega\mu\kappa_\varepsilon \tag{2.1.38}$$

with an a.c. conductivity

$$\kappa_\varepsilon = \kappa + \omega\varepsilon'' \tag{2.1.39}$$

Comparing Equation 2.1.38 with Equation 2.1.26, we arrive at the following analogy:

- Permittivity $\varepsilon \Leftrightarrow \varepsilon'$ real part of the complex permittivity.
- Conductivity $\kappa \Leftrightarrow \kappa_\varepsilon$ a.c. conductivity.

If we use a.c. conductivity κ_ε in Equations 2.1.27 and 2.1.28, for instance, instead of conductivity κ, losses can be easily included.

Furthermore, from Equations 2.1.11 and 2.1.22, we get the relationship

$$\underline{\beta} = k - j\frac{\alpha}{2} = \frac{2\pi n}{\lambda_0} - j\frac{\alpha}{2} = \frac{2\pi}{\lambda_0}\left(n - j\frac{\alpha\lambda_0}{4\pi}\right) \tag{2.1.40}$$

or

$$\underline{\beta} = \frac{2\pi}{\lambda_0}(n - jn') \tag{2.1.41}$$

Consequently, it applies for the extinction coefficient that

$$n' = \frac{\alpha\lambda_0}{4\pi} \tag{2.1.42}$$

Once more it becomes clear that the electric and optical characteristics of a material are interrelated. It is therefore possible to experimentally determine a material's optical parameters by measuring permeability μ, permittivity ε and conductivity κ.

Example 2.1: Power Dissipation in Dielectrics

For effective power P_V transformed in volume V_D it applies that

$$P_V = \frac{\tilde{V}^2}{V_D R} \tag{2.1.43}$$

with effective value of the voltage $\tilde{V}$ and real part R of complex resistivity $\underline{Z}$

$$R = \mathrm{Re}\{\underline{Z}\} = \mathrm{Re}\left\{\frac{1}{j\omega\underline{C}}\right\} \tag{2.1.44}$$

As the capacity is lossy, it is also assumed to be complex:

$$\underline{C} = \frac{\underline{\varepsilon}A}{d} \tag{2.1.45}$$

with surface A and height d of volume V_D. The field strength in the volume results from the applied voltage

$$\tilde{V} = \frac{1}{\sqrt{2}}Ed \tag{2.1.46}$$

It results that:

$$P_V = \frac{1}{2}E^2\,\mathrm{Re}\,\{j\omega\underline{\varepsilon}\} \tag{2.1.47}$$

We can arrive at complex permeability $\underline{\varepsilon}$ by using complex susceptibility $\underline{\chi}$:

$$\underline{\varepsilon} = \varepsilon_0(1 + \underline{\chi}) \tag{2.1.48}$$

and

$$\underline{\chi} = \chi' - j\chi'' = \chi(\cos\theta - j\sin\theta) \tag{2.1.49}$$

with

$$\tan\theta = \frac{\chi''}{\chi'} \tag{2.1.50}$$

Here, angle θ corresponds to the loss angle or loss factor $\tan\delta$, respectively:

$$\tan\delta = \frac{\varepsilon''}{\varepsilon'} = \frac{1 + \chi''}{1 + \chi'} \tag{2.1.51}$$

Thus, the real part of Equation 2.1.47 becomes

$$\mathrm{Re}\{j\omega\,\underline{\varepsilon}\} = \omega\varepsilon_0\chi\sin\theta \tag{2.1.52}$$

The volume-related power dissipation in the dielectric then becomes

$$P_V = \frac{1}{2}E^2\omega\varepsilon_0\,\chi\sin\theta \tag{2.1.53}$$

2.1.3   Fields at Interfaces

If an electromagnetic wave impinges on an interface, part of the radiation passes through the interface, that is it is transmitted, whereas another part of the radiation is reflected (Figure 2.1.1). The incoming wave – marked by index i – propagates with refractive index n_i in direction of wave vector $\vec{k}_i$ through the isotrope medium. Vector $\vec{E}_i$ of the electric field strength stands perpendicular to the page in Figure 2.1.1 and is directed into the page. The incident power of the radiation is

$$P_i = I_i A \cos\varphi_i \tag{2.1.54}$$

with beam cross-section $A\cos\varphi_i$. Part of the radiation is reflected at the interface, and another part enters the medium (refractive index n_t). Reflected power P_r or transmitted power P_t, respectively, is

$$P_r = I_r A \cos\varphi_r \tag{2.1.55}$$

$$P_t = I_t A \cos\varphi_t \tag{2.1.56}$$

with beam cross-sections $A\cos\varphi_i$ and $A\cos\varphi_t$.

It results for reflectance ρ

$$\rho = r^2 = \frac{P_r}{P_i} = \frac{I_r}{I_i} = \frac{E_{or}^2}{E_{0i}^2} \tag{2.1.57}$$

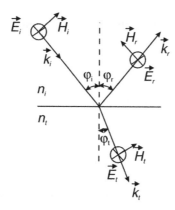

Figure 2.1.1   Refraction at the interface of media with optically different densities (refractive indices n_i, n_t)

Here, the law of reflection was taken into consideration:

$$\varphi_i = \varphi_r \tag{2.1.58}$$

Analogously, it applies for transmittance τ that

$$\tau = \frac{n_t \cos \varphi_t}{n_i \cos \varphi_i} \frac{E_{0t}^2}{E_{0i}^2} = \frac{n_t \cos \varphi_t}{n_i \cos \varphi_i} t^2 \tag{2.1.59}$$

SNELL's law of refraction applies for the angle of the transmitted wave:

$$n_i \cos \varphi_i = n_t \cos \varphi_t \tag{2.1.60}$$

If there is no absorption, it has to apply due to energetic reasons that:

$$\rho + \tau = 1 \tag{2.1.61}$$

2.1.4   Transmission Through Thin Dielectric Layers

We talk about thin dielectric layers if the layer's thickness lies in the same order as the wavelength of the radiation, that is in the nano- or micrometre range. This will cause interferences in the layers due to the superposition of reflected, transmitted and incident waves. The result is that the layer's transmission and absorption does not only depend on the refractive index, but also on the layer thickness. The calculation uses the method of transfer matrices.

For a linearly polarised wave that impinges on thin dielectric layers, [2] applies:

$$\begin{bmatrix} E_I \\ H_I \end{bmatrix} = \mathbf{M} \begin{bmatrix} E_{II} \\ H_{II} \end{bmatrix} \tag{2.1.62}$$

with electric and magnetic field strength components at the first layer (index I) and second layer (index II) and transfer matrix **M**:

$$\mathbf{M} = \begin{pmatrix} m_{11} & m_{12} \\ m_{21} & m_{22} \end{pmatrix} \qquad (2.1.63)$$

Calculating the reflectance ρ and transmittance τ, we use a simplification and assume a perpendicular incidence of the radiation[2] [2]:

$$\rho = \left| \frac{G_{I}m_{11} + G_{I}G_{II}m_{12} - m_{21} - G_{II}m_{22}}{G_{I}m_{11} + G_{I}G_{II}m_{12} + m_{21} + G_{II}m_{22}} \right|^{2} \qquad (2.1.64)$$

$$\tau = \left| \frac{2G_{I}}{G_{I}m_{11} + G_{I}G_{II}m_{12} + m_{21} + G_{II}m_{22}} \right|^{2} \qquad (2.1.65)$$

with

$$G_{I} = n_{I}\sqrt{\frac{\varepsilon_{0}}{\mu_{0}}} \qquad (2.1.66)$$

and

$$G_{II} = n_{II}\sqrt{\frac{\varepsilon_{0}}{\mu_{0}}} \qquad (2.1.67)$$

If we have a layer system with i layers, for example when infrared radiation enters a sensor volume through the air via an antireflection layer ($i = 3$), transfer matrix **M** will be the product of the transfer matrices $\mathbf{M}_i$ of the individual layers i:

$$\mathbf{M} = \prod_{i} \mathbf{M}_{i} \qquad (2.1.68)$$

where it applies, for each layer, that:

$$\mathbf{M}_{i} = \begin{pmatrix} \cos\gamma_{i} & j\dfrac{\sin\gamma_{i}}{G_{i}} \\ jG_{i}\sin\gamma_{i} & \cos\gamma_{i} \end{pmatrix} \qquad (2.1.69)$$

with

$$\gamma_{i} = \frac{2\pi}{\lambda}n_{i}d_{i} \qquad (2.1.70)$$

and

$$G_{i} = n_{i}\sqrt{\frac{\varepsilon_{0}}{\mu_{0}}}. \qquad (2.1.71)$$

Each individual transfer matrix $\mathbf{M}_i$ characterises the transmission of the radiation from layer i (thickness d_i and refractive index n_i) into layer ($i + 1$). It is important for the calculation to follow the order of the layers, as the ray moves through the layers with increasing indices.

---

[2] This is the technically most important case and allows an easy-to-handle formula.

Example 2.2: Absorption in a Microbolometer Bridge

The active part of a microbolometer bridge consists of a system of thin layers that is situated about 2.5 μm above a mirror (Figures 6.6.3 and 6.6.4). The microbolometer bridge works in vacuum. The radiant flux comes from the vacuum and moves through the layer system of the microbolometer bridge. The non-absorbed part passes through the bridge, exits to the vacuum underneath the bridge, is reflected at the mirror on the chip and returns to the bolometer bridge. Bridge and mirror form a $\lambda/4$-resonator that absorbs maximum radiation.

a. Transmission of the Layer System

The absorption in the layer system can be neglected due to the very small layer thickness according to term $\exp\left\{-\dfrac{\alpha}{2}z\right\}$ of Equation 2.1.10. Transmittance τ_B is then calculated using Equation 2.1.64. Table 2.1.2 presents the corresponding refractive indices n_i. Figure 2.1.2 shows the calculated transmission.

Table 2.1.2 Refractive indices n for thin layers in the microbolometer pixel

Material	Refractive index n
SiO$_2$	1.47
a-Si	4.0
Si$_3$N$_4$	2.0
VOx	3.3
TiN	1.6
TiO$_2$	2.12
Vacuum	1.00

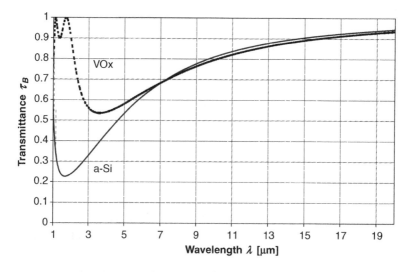

Figure 2.1.2   Wavelength-dependent transmission τ_B of a microbolometer bridge

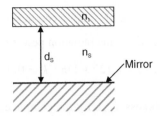

Figure 2.1.3   Layer structure of a $\lambda/4$-resonator

b. Absorption in the Resonator

For absorbing the radiation, we use a $\lambda/4$-resonator (Figure 2.1.3). It consists of a mirrored floor beneath the microbridge, layer S with refractive index n_S (vacuum) and a bolometer layer with refractive index n_1, where it applies that:

$$n_1 \neq n_S \tag{2.1.72}$$

The principle is based on the superposition of two electromagnetic waves with intensities I_1 and I_2

$$I = I_1 + I_2 + 2\sqrt{I_1 I_2} \cos \delta \tag{2.1.73}$$

and a phase difference of

$$\delta = \frac{2\pi}{\lambda} \Delta \tag{2.1.74}$$

Here, Δ is the path difference expressed as the difference of the optical thickness (refractive index multiplied by path) for both waves. For the complete reflection of perpendicularly incoming beams, it applies that:

$$I_1 = I_2 = I_0 \tag{2.1.75}$$

The beam moves forth and back in layer S with optical thickness $(n_S \cdot d_S)$, which results in a path difference of

$$\Delta = 2n_S d_S \tag{2.1.76}$$

The intensity of the reflected wave now becomes:

$$I = 2I_0 \left(1 + \cos \frac{4\pi n_S}{\lambda} d_S \right) \tag{2.1.77}$$

For energetic reason – if there is no transmission ($\tau_S = 0$) – the sum of reflected and absorbed radiation has to be constant:

$$\rho_S + \alpha_S = 1 \tag{2.1.78}$$

with ρ_S as the reflectance and α_S as the absorption coefficient. For destructive interference ($I=0$, no reflection), the absorption of the radiation power occurs for $\cos\dfrac{4\pi\,n_S}{\lambda}d_S = -1$:

$$\frac{4\pi n_s}{\lambda}d_{S,\lambda/4} = (2k+1)\pi \quad k = 0,1\ldots \tag{2.1.79}$$

For a given wavelength, thickness $d_{S,\lambda/4}$ for the complete absorption is calculated as

$$d_{S,\lambda/4} = \frac{2k+1}{n_s}\frac{\lambda}{4} \tag{2.1.80}$$

To calculate the absorption coefficient, we normalise Equation 2.1.77 to the maximum intensity

$$I_{\max} = 4I_0 \tag{2.1.81}$$

Taking into consideration Equation 2.1.57, it results for the absorption coefficient that

$$\alpha_{\lambda/4} = 1 - \frac{I}{I_{\max}} = \frac{1}{2}\left(1 - \cos\frac{4\pi n_S}{\lambda}d_S\right) \tag{2.1.82}$$

Figure 2.1.4 shows an example.

c. Total Absorption

Total absorption α is determined by the transmission of layer system τ_B and absorption $\alpha_{\lambda/4}$ of the $\lambda/4$-resonator:

$$\alpha = \tau_B\alpha_{\lambda/4} \tag{2.1.83}$$

Figure 2.1.5 presents the total absorption of the selected example parameters in Figures 2.1.3 and 2.1.4. For commercial microbolometers, the absorption by transmission of the sensor window (optical bandpass) is additionally limited to the technically desirable range of between approximately 8 and 14 µm.

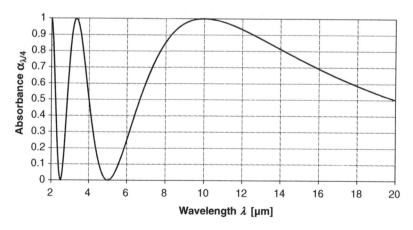

Figure 2.1.4 Absorption coefficient $\alpha_{\lambda/4}$ of a $\lambda/4$-resonator. Layer S: $n_S = 1$ (vacuum), $d_S = 2.5\,\mu m$

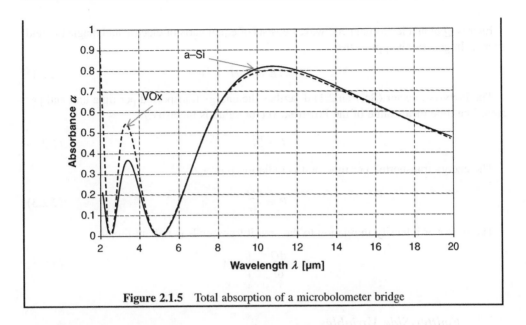

Figure 2.1.5   Total absorption of a microbolometer bridge

2.2   Radiation Variables

Thermal sensors quantitatively determine the intensity of electromagnetic radiation. This process is also called radiometric measurement. In the following, we will describe the different radiation-physical variables that can be used to characterise radiation (Table 2.2.1).

2.2.1   Radiation-Field-Related Variables

Radiation-field-related variables in general describe the electromagnetic radiation field or the electromagnetic wave in space.

Table 2.2.1   Radiometric radiation variables

Variable	Denomination	Symbol	Unit
Radiation-field-related	Radiation energy	Q	Ws
	Radiant intensity	I	W/m^2
	Radiation power	P	W
	Radiant flux	Φ	W
Emitter-related	Radiant intensity	I	W/sr
	Radiant exitance	M_S	W/m^2
	Radiance	L	$W/(sr\ m^2)$
Receiver-related	Irradiance	E	W/m^2
	Irradiation	H	$Ws/m^2 = J/m^2$
Spectral	Spectral intensity	I_λ	$W/(sr\ \mu m)$
	Spectral radiance	L_λ	$W/(sr\ m^2\ \mu m)$
	Spectral radiant exitance	$M_{\lambda S}$	$W/(m^2\ \mu m)$

Energy Q of an electromagnetic wave consists of equal parts of electric and magnetic field energy. In general, it applies that

$$Q = \varepsilon E^2 = \mu H^2 \tag{2.2.1}$$

The POYNTING vector (Section 2.1) describes the energy transported per time unit and per area. The time mean value of the POYNTING vector value is intensity I:

$$I = \overline{|\vec{S}|} \tag{2.2.2}$$

The energy transported per time unit is radiation power P:

$$P = \frac{dQ}{dt} \tag{2.2.3}$$

The power or intensity in relation to the area it transcends is radiant flux Φ:

$$\Phi = \frac{dQ}{dt} = \int_A I \, dA \tag{2.2.4}$$

2.2.2 Emitter-Side Variables

The emitter-related variables describe the radiometric radiation variables from a sender's or emitter's perspective.

Radiant intensity I is the radiant flux per solid angle unit:

$$I = \frac{d\Phi}{d\Omega} \tag{2.2.5}$$

The radiant flux in relation to an area is exitance M_{BB}:

$$M_{BB} = \frac{d\Phi}{dA} \tag{2.2.6}$$

Index BB indicates that the exitance refers to a blackbody (Example 2.3). Radiance L is the radiant flux per area unit in normal direction and per solid angle unit:

$$L = \frac{d^2\Phi}{dA \cos \gamma \, d\Omega} \tag{2.2.7}$$

with angle γ between surface normal and solid-angle element.

2.2.3 Receiver-Related Variables

The receiver-related variables describe the radiometric radiation variable from a sensor's and receiving surface's perspective.

Irradiance E is the radiant flux per area unit of the irradiated surface:

$$E = \frac{d\Phi}{dA} \tag{2.2.8}$$

Irradiation is a time integral of the irradiance:

$$H = \int E \, dt \tag{2.2.9}$$

2.2.4 Spectral Variables

Spectral variables refer to wavelength interval $d\lambda$ and are marked by index λ. All radiation-field-, emitter- and receiver-related variables can be stated as spectral variables. The most important ones are:

$$\text{Spectral intensity} \quad I_\lambda = \frac{d^2 \Phi}{d\Omega \, d\lambda} \tag{2.2.10}$$

$$\text{Spectral radiance} \quad L_\lambda = \frac{d^3 \phi}{dA \cos \varphi d\Omega d\lambda} \tag{2.2.11}$$

$$\text{Spectral exitance} \quad M_{\lambda BB} = \frac{d^2 \phi}{dA \, d\lambda} \tag{2.2.12}$$

Spectral variables can also be related to frequency interval dv (marked with index v). As frequency and wavelength are not linearly related

$$v = \frac{c}{\lambda} \tag{2.2.13}$$

intervals $d\lambda$ and dv are not the same either. With

$$\frac{dv}{d\lambda} = \frac{d\left(\frac{c}{\lambda}\right)}{d\lambda} = -\frac{c}{\lambda^2} \tag{2.2.14}$$

it follows that[3]

$$dv = \frac{c}{\lambda^2} d\lambda \tag{2.2.15}$$

As it has to apply that

$$L = \int_0^\infty L_\lambda(\lambda) d\lambda = \int_0^\infty L_v(v) dv \tag{2.2.16}$$

it also applies that

$$L_\lambda(\lambda) d\lambda = L_v(v) dv \tag{2.2.17}$$

---

[3] The minus sign in Equation 2.2.14 means that an increase in dv results in a decreased $d\lambda$. This is neglegible for the conversion of the intervals, and in the following, we will only consider the absolute value.

It follows with Equation 2.2.14:

$$L_\lambda(\lambda) = \frac{c}{\lambda^2} L_\nu(\nu) \qquad (2.2.18)$$

2.2.5 Absorption, Reflection and Transmission

If radiation Φ_0 transcends a body, it can be absorbed (portion Φ_a), reflected (portion Φ_r) or transmitted (portion Φ_t) (Figure 2.2.1). In order to determine the individual portions, normalised variables are defined for incoming radiant flux Φ_0. The wavelength-related absorption is described by the spectral absorption coefficient

$$\alpha_\lambda = \frac{\Phi_a(\lambda)}{\Phi_0(\lambda)} \qquad (2.2.19)$$

If the absorption coefficient refers to a wavelength range of λ_1 to λ_2, we call it the band absorption coefficient:

$$\alpha_B = \frac{\displaystyle\int_{\lambda_1}^{\lambda_2} \Phi_a(\lambda)d\lambda}{\displaystyle\int_{\lambda_1}^{\lambda_2} \Phi_0(\lambda)d\lambda} \qquad (2.2.20)$$

If we look at the entire wavelength range of $\lambda_1 = 0$ to $\lambda_2 = \infty$ we call it integral or total absorption:

$$\alpha = \frac{\displaystyle\int_0^\infty \Phi_a(\lambda)d\lambda}{\displaystyle\int_0^\infty \Phi_0(\lambda)d\lambda} \qquad (2.2.21)$$

Correspondingly, we also define reflection coefficients and transmission coefficients. It applies for the spectral reflection coefficient that:

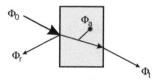

Figure 2.2.1 Absorption, reflection and transmission at a body

$$\rho_\lambda = \frac{\Phi_r(\lambda)}{\Phi_0(\lambda)} \tag{2.2.22}$$

The band reflection coefficient is

$$\rho_B = \frac{\displaystyle\int_{\lambda_1}^{\lambda_2} \Phi_r(\lambda)\,d\lambda}{\displaystyle\int_{\lambda_1}^{\lambda_2} \Phi_0(\lambda)\,d\lambda} \tag{2.2.23}$$

and the integral reflection coefficient is

$$\rho = \frac{\displaystyle\int_0^\infty \Phi_r(\lambda)\,d\lambda}{\displaystyle\int_0^\infty \Phi_0(\lambda)\,d\lambda} \tag{2.2.24}$$

For the spectral transmission coefficient, it applies that:

$$\tau_\lambda = \frac{\Phi_t(\lambda)}{\Phi_0(\lambda)} \tag{2.2.25}$$

The band transmission coefficient is

$$\tau_B = \frac{\displaystyle\int_{\lambda_1}^{\lambda_2} \Phi_t(\lambda)\,d\lambda}{\displaystyle\int_{\lambda_1}^{\lambda_2} \Phi_0(\lambda)\,d\lambda} \tag{2.2.26}$$

and the integral transmission coefficient

$$\tau = \frac{\displaystyle\int_0^\infty \Phi_t(\lambda)\,d\lambda}{\displaystyle\int_0^\infty \Phi_0(\lambda)\,d\lambda} \tag{2.2.27}$$

Absorption, reflection and transmission coefficients of a body depend, amongst other things, on the body's temperature, chemical composition, surface characteristics, as well as on the polarisation of the radiation. In general, they have to be determined by measurements. Often they also depend on time, for example when a body's surface oxidises. From the principle of the conservation of energy, it results that

$$\alpha_\lambda + \rho_\lambda + \tau_\lambda = 1 \qquad (2.2.28)$$

and

$$\alpha + \rho + \tau = 1 \qquad (2.2.29)$$

For impermeable bodies ($\tau_\lambda = 0$), it thus applies that

$$\alpha_\lambda = 1 - \rho_\lambda \qquad (2.2.30)$$

2.2.6 Emissivity

Spectral emissivity is the relation of spectral exitance M_λ of a body to that of a blackbody $M_{\lambda BB}$ at a certain temperature T and wavelength λ:

$$\varepsilon_\lambda(T) = \frac{M_\lambda(\lambda, T)}{M_{\lambda BB}(\lambda, T)} \qquad (2.2.31)$$

Spectral emissivity can also be related to the direction and be stated as radiance ratio:

$$\varepsilon_\lambda(T, \vartheta, \varphi) = \frac{L_\lambda(\lambda, T, \vartheta, \varphi)}{L_{\lambda BB}(\lambda, T)} \qquad (2.2.32)$$

with radiance L_λ of a body in azimuth ϑ and evaluation angle φ and radiance $L_{\lambda BB}$ of a blackbody. The emission coefficient can also be stated as band emission coefficient

$$\varepsilon_B(T, \vartheta, \varphi) = \frac{\displaystyle\int_{\lambda_1}^{\lambda_2} L_\lambda(T, \lambda, \vartheta, \varphi)\,d\lambda}{\displaystyle\int_{\lambda_1}^{\lambda_2} L_{\lambda BB}(T, \lambda)\,d\lambda} \qquad (2.2.33)$$

and total emission coefficient

$$\varepsilon(T, \vartheta, \varphi) = \frac{\displaystyle\int_0^\infty L_\lambda(T, \lambda, \vartheta, \varphi)\,d\lambda}{\displaystyle\int_0^\infty L_{\lambda BB}(T, \lambda)\,d\lambda} = \frac{\displaystyle\int_0^\infty L_\lambda(T, \lambda, \vartheta, \varphi)\,d\lambda}{\dfrac{\sigma T^4}{\pi}} \qquad (2.2.34)$$

Example 2.3: Blackbody

The so-called blackbody is of outstanding importance. It is the ideal radiation source for thermal radiation and emits for each wavelength the maximum possible thermal radiation. Its spectral radiance $L_{\lambda BB}$ corresponds exactly to PLANCK'S radiation law (Section 2.3):

$$L_{\lambda BB} = \frac{M_{\lambda BB}}{\pi} = \frac{c_1}{\pi \lambda^5} \frac{1}{e^{\frac{c_2}{\lambda T}} - 1} \tag{2.2.35}$$

with radiation constants c_1 and c_2. It has a wavelength-independent emissivity and absorption coefficients of unity and is a LAMBERTian radiator. LAMBERTian radiators have a direction-independent radiance (Section 3.2):

$$L_S(\vartheta, \varphi) = \text{const.} \tag{2.2.36}$$

It results for the intensity that

$$I(\vartheta) = I_0 \cos \vartheta \tag{2.2.37}$$

with intensity I_0 in the normal direction. This means that a blackbody has always the same 'brightness', independent of the line of vision towards the blackbody. The STEFAN–BOLTZMANN law (Section 2.3) can now be applied to calculate its total radiation:

$$L_S = \int_0^\infty L_\lambda(T) d\lambda = \frac{\sigma T^4}{\pi} \tag{2.2.38}$$

Due to these characteristics, it is called a 'black emitter' or also a 'blackbody'. It does not reflect any radiation and, at an emitter temperature below 400 °C, its background radiation in the visible wavelength range becomes too small to be visible. Therefore, its surface appears matt black.

As it is difficult to technically realise an emissivity of unity,[4] often 'grey bodies' are used. They have the same characteristics as blackbodies, with the exception of the emissivity being smaller than 1.

2.3 Radiation Laws

Table 2.3.1 summarizes the most important radiation laws. Section 3.2 describes the photometric law in detail. Appendix B provides the derivation of PLANCK'S radiation law and the laws derived from it.

PLANCK'S radiation law describes the spectral radiant exitance $M_{\lambda BB}$ of a blackbody in the wavelength range $\lambda + d\lambda$ of a differential area dA into the half-space. Figures 2.3.1a and 2.3.2b

---

[4] Due to the statistic characteristics of emission and absorption, it is impossible to achieve an emissivity of exactly 1.

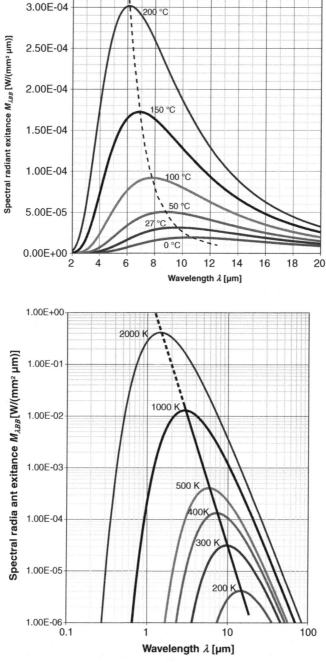

Figure 2.3.1 PLANCK'S radiation law (a) in linear and (b) in double-logarithmic representation Parameter: temperature. The dashed line marks WIEN'S displacement law.

Table 2.3.1 Radiation laws

Name	Formula
Photometric law	$$d^2\Phi_{12} = L_1 \frac{dA_1 \cos \beta_1 dA_2 \cos \beta_2}{r^2} \quad (2.3.1)$$
PLANCK'S radiation law	$$M_{\nu BB} = 2\pi \frac{h\nu^3}{c^2} \frac{1}{e^{\frac{h\nu}{k_B T}} - 1} \quad (2.3.2)$$
	$$M_{\lambda BB} = \frac{c_1}{\lambda^5} \frac{1}{e^{\frac{c_2}{\lambda T}} - 1} \quad (2.3.3)$$
Stefan–Boltzmann law	$$M_{BB} = \sigma T^4 \quad (2.3.4)$$
WIEN'S displacement law	$$\nu_{max} = 5.878 \times 10^{10} \frac{\text{Hz}}{\text{K}} T \quad (2.3.5)$$
	$$\lambda_{max} = \frac{2.898 \times 10^{-3}\text{K m}}{T} \quad (2.3.6)$$

$d^2\Phi_{12}$ radiant flux originating from area element dA_1 and received by area element dA_2; r distance between area elements dA_1 and dA_2; β_1, β_2 angles of area elements dA_1 and dA_2 in relation to r; L_1 radiance of area element dA_1.

provide a linear and a double-logarithmic representation of the radiant exitance, respectively. Figure 2.3.2 shows an unusual representation of the exitance over the frequency. Figures 2.3.1 and 2.3.2 marks the wavelength or frequency where the exitance reaches its maximum (WIEN'S displacement law).

Integrating PLANCK'S radiation law and taking into consideration spectral emissivity ε_λ, we arrive at exitance M of any emitter in a defined wavelength range λ_1 to λ_2 (Figure 2.3.3 and Table 2.3.2):

$$M = \int_{\lambda_1}^{\lambda_2} \varepsilon_\lambda \frac{c_1}{\lambda^5} \frac{1}{e^{\frac{c_2}{\lambda T}} - 1} d\lambda \quad (2.3.7)$$

This integral can be numerically solved for any wavelength range. There is a single, known self-contained solution for total radiation ($\lambda_1 = 0$ to $\lambda_2 = \infty$) and for greybodies (STEFAN–BOLTZMANN law):

$$M = \varepsilon \sigma T^4 \quad (2.3.8)$$

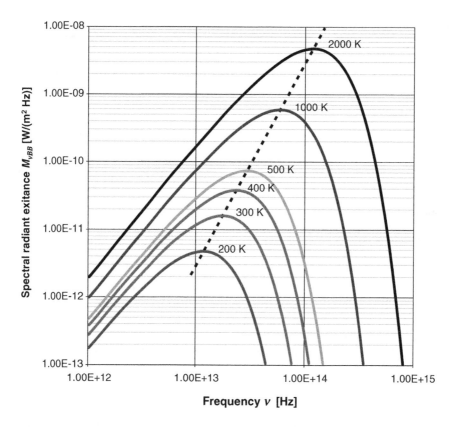

Figure 2.3.2 PLANCK's radiation law versus frequency v. Parameter: temperature. The dashed line marks WIEN's displacement law

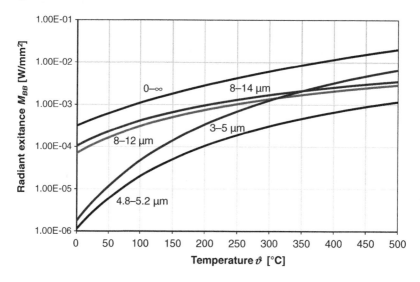

Figure 2.3.3 Exitance M_{BB} of blackbodies in different wavelength ranges

Table 2.3.2   Values of exitance M_{BB} of blackbodies at different wavelength ranges

Temperature ϑ [°C]	M_{BB} [W/mm²]				
	0–∞	3–5 µm	4.8–5.2 µm	8–12 µm	8–14 µm
0	3.16×10^{-4}	1.76×10^{-6}	1.13×10^{-6}	7.03×10^{-5}	1.04×10^{-4}
10	3.64×10^{-4}	2.70×10^{-6}	1.65×10^{-6}	8.53×10^{-5}	1.25×10^{-4}
20	4.19×10^{-4}	4.03×10^{-6}	2.36×10^{-6}	1.02×10^{-4}	1.47×10^{-4}
30	4.79×10^{-4}	5.86×10^{-6}	3.28×10^{-6}	1.21×10^{-4}	1.73×10^{-4}
50	6.18×10^{-4}	1.16×10^{-5}	5.96×10^{-6}	1.65×10^{-4}	2.30×10^{-4}
80	8.82×10^{-4}	2.82×10^{-5}	1.29×10^{-5}	2.45×10^{-4}	3.34×10^{-4}
100	1.10×10^{-3}	4.73×10^{-5}	2.01×10^{-5}	3.09×10^{-4}	4.16×10^{-4}
200	2.84×10^{-3}	3.37×10^{-4}	1.05×10^{-4}	7.45×10^{-4}	9.58×10^{-4}
300	6.12×10^{-3}	1.26×10^{-3}	3.09×10^{-4}	1.35×10^{-3}	1.69×10^{-3}
400	1.16×10^{-2}	3.23×10^{-3}	6.62×10^{-4}	2.08×10^{-3}	2.56×10^{-3}
500	2.03×10^{-2}	6.60×10^{-3}	1.17×10^{-3}	2.90×10^{-3}	3.54×10^{-3}

Example 2.4: Exitance Curve

We will analyse what portion of the exitance is situated in the wavelength range $\lambda = 0$ to λ_{max}. This question is for instance relevant if we want to calculate the portion of visible light or the effective IR radiation in a wavelength-limited sensor element. It applies to the ratio R between detected and total exitance that

$$R = \frac{\displaystyle\int_0^{\lambda_{max}} \frac{c_1}{\lambda^5} \frac{1}{e^{\frac{c_2}{\lambda T}}-1} \, d\lambda}{\displaystyle\int_0^{\infty} \frac{c_1}{\lambda^5} \frac{1}{e^{\frac{c_2}{\lambda T}}-1} \, d\lambda} \tag{2.3.9}$$

The solution of the denominator in Equation 2.3.9 corresponds to the STEFAN–BOLTZMANN law according to Equation 2.3.8. We can only approximately calculate the numerator. Substituting

$$x = \frac{c_2}{\lambda T} \tag{2.3.10}$$

and using

$$d\lambda = -\frac{c_2}{\lambda^2 T} \, dx \tag{2.3.11}$$

it follows for the numerator integral that

$$\int_0^{\lambda_{max}} \frac{c_1}{\lambda^5} \frac{1}{e^{\frac{c_2}{\lambda T}}-1} \, d\lambda = -\frac{c_1 T^4}{c_2^4} \int_{x_1}^{x_2} \frac{x^3}{e^x-1} \, dx \tag{2.3.12}$$

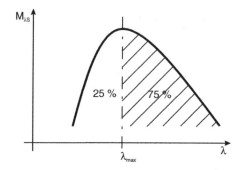

Figure 2.3.4   Relation of emitted power to the wavelength of maximum exitance λ_{max}

For the limits of the integral, it applies that

$$x_1 = \infty \tag{2.3.13}$$

and with Equation B.62

$$x_2 = \frac{c_2}{\lambda_{max}T} = 4.965 \tag{2.3.14}$$

Applying Equation 2.3.10 and the STEFAN–BOLTZMANN law in Equations 2.3.8 and 2.3.9 becomes

$$R = \frac{15}{\pi^4} \int\limits_{x_2}^{x_1} \frac{x^3}{e^x - 1} dx \tag{2.3.15}$$

As x is always larger than x_2, we can neglect 1 in the denominator:

$$\int\limits_{x_2}^{x_1} \frac{x^3}{e^x - 1} dx \approx \int\limits_{x_2}^{x_1} x^3 e^{-x} dx \tag{2.3.16}$$

The resulting deviation is thus smaller than 1%. The solution of the integral in Equation 2.3.16 can be found in suitable integral tables [3]:

$$\int x^3 e^{-x} dx = -e^{-x}[x^3 + 3x^2 + 6x + 6] \tag{2.3.17}$$

For the upper limit $x_1 = \infty$, there results an integral value of zero. And thus the result becomes

$$R = 0.250 \tag{2.3.18}$$

This means that 25% of the emitted power is radiated below and 75% of the power is radiated above the maximum wavelength (Figure 2.3.4). This value is independent of the emitter's temperature!

References

1. Glaser, W. (1997) *Photonik für Ingenieure (Photonics for Engineers)*, Verlag Technik, Berlin.
2. Hecht, E. (2003) *Optics*, Addison-Wesley, Reading.
3. Bronstein, I.N., Semendyayev, K.A., Musiol, G. and Muehlig, H. (2002) *Handbook of Mathematics*, Springer-Verlag, Berlin.

3

Photometric Basics

The incoming signal of a thermal infrared sensor is the radiation emitted by a body. The radiation power that the sensor receives depends on the optical-geometric imaging relations. In the following, we will present the solid angle and the photometric law that describe these imaging relations.

3.1 Solid Angle

3.1.1 Definition

The solid angle provides an exact description of a point source and is thus a suitable description of the propagation of radiation. It is defined as the relation of sphere surface A_S that limits the propagation space and the squared radius r of this sphere (Figure 3.1.1):

$$\Omega = \frac{A_S}{r^2}\Omega_0 \tag{3.1.1}$$

with solid angle unit steradian

$$\Omega_0 = 1 \text{ sr} = \frac{1 \text{m}^2}{1 \text{m}^2} \tag{3.1.2}$$

One steradian cuts out an area of 1 m^2 off the unit sphere ($r = 1$ m).[1] The area's form can be arbitrary. Identical surface areas with different forms always have the same solid angle. The unit of solid angle (sr or Ω_0) is 1 and could in principle be dropped. However, it appears reasonable to provide the unit in order to keep the physical references.

Applying definition Equation 3.1.1, the maximum solid angle Ω_{max} for a full sphere can be calculated using

$$A_{Sfullsphere} = 4\pi r^2 \tag{3.1.3}$$

---

[1] This definition is modelled based on the plane angle with the unit radian: $1 \text{ rad} = \frac{1\text{m}}{1\text{m}}$. One radian corresponds to 1 m of circumference of the unit circle ($r = 1$ m).

Thermal Infrared Sensors: Theory, Optimisation and Practice   Helmut Budzier and Gerald Gerlach
© 2011 John Wiley & Sons, Ltd

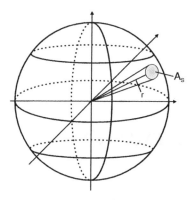

Figure 3.1.1   Solid angle definition

Thus it becomes

$$\Omega_{max} = 4\pi sr = 4\pi\,\Omega_0 \qquad\qquad (3.1.4)$$

and the solid angle for the half-space becomes:

$$\Omega_{HS} = 2\pi\,\Omega_0 \qquad\qquad (3.1.5)$$

3.1.2   Solid Angle Calculations

3.1.2.1 Right Circular Cone

A right circular cone cuts out a spherical cap with height h and area A_{CC} off the sphere (Figure 3.1.2). Solid angle Ω_{CC} of the right circular cone is calculated as:

$$A_{CC} = 2\pi r_0 h \qquad\qquad (3.1.6)$$

$$h = r_0(1-\cos\,\varphi_C) \qquad\qquad (3.1.7)$$

$$\Omega_{CC} = 2\pi(1-\cos\,\varphi_C)\Omega_0 \qquad\qquad (3.1.8)$$

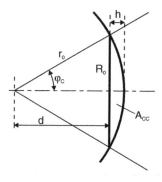

Figure 3.1.2   Right circular cone

or

$$\Omega_{CC} = 4\pi \sin^2 \frac{\varphi_C}{2} \Omega_0 \qquad (3.1.9)$$

The result for a solid angle Ω_{CC}, for instance, for a field of view of 120° ($\varphi_C = 60°$) is

$$\Omega_{120°} = \pi \Omega_0 \qquad (3.1.10)$$

For smaller angles, it applies that $\sin \varphi_C \approx \varphi_C$:

$$\Omega_{CC} \approx \pi \varphi_C^2 \Omega_0 \qquad (3.1.11)$$

Substituting

$$\cos \varphi_C = \sqrt{1 - \sin^2 \varphi_C} \qquad (3.1.12)$$

and

$$\sin \varphi_C = \frac{R_0}{r_0} \qquad (3.1.13)$$

in Equation 3.1.8, the result is the solid angle of the right circular cone in the form of

$$\Omega_{CC} = 2\pi \left[1 - \sqrt{1 - \left(\frac{R_0}{r_0}\right)^2} \right] \qquad (3.1.14)$$

The solid angle of a right circular cone is also called canonical solid angle.

Many calculations require differential solid angle $\frac{d\Omega_{CC}}{d\varphi_C}$, which is calculated using Equation 3.1.8:

$$d\Omega_{CC} = 2\pi \sin \varphi_C \, d\varphi_C \, \Omega_0 \qquad (3.1.15)$$

3.1.2.2 Arbitrary Areas

Solid angle Ω of any area A situated in space is the area that results from central projection of A to the unit sphere, divided by the unit radius (Figure 3.1.3):

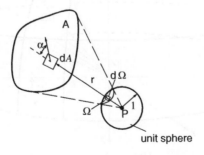

Figure 3.1.3 Solid angle of any area

$$\frac{dA \cos \alpha}{d\Omega} = \frac{r^2}{1^2} \tag{3.1.16}$$

or

$$d\Omega = \frac{dA \cos \alpha}{r^2} \tag{3.1.17}$$

It applies for the solid angle.

- in the spherical coordinate system that

$$\Omega = \int_A \frac{\cos \alpha}{r^2} \, dA \tag{3.1.18}$$

- in the Cartesian coordinate system that

$$\Omega = \int\int \frac{\cos \alpha}{r^2(x,y)} \, dx \, dy \tag{3.1.19}$$

- and in the polar coordinate system that

$$\Omega = \int\int \sin \gamma \, d\gamma \, d\varphi \tag{3.1.20}$$

Here, angle γ is the angle of the area element in relation to the positive z-axis and angle φ is the angle of the area element in relation to the positive x-axis (Figure 3.1.4).

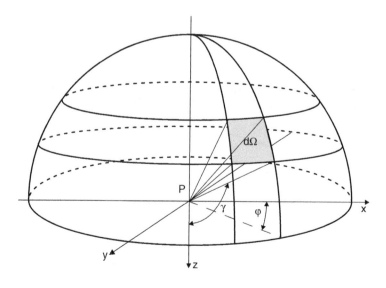

Figure 3.1.4 Solid angle in polar coordinates

Figure 3.1.5   Vectors for any origin of coordinates

It is advantageous to state the solid angle as vectors. For this reason, we introduce r into Equation 3.1.18:

$$\Omega = \int_A \frac{r \cos \alpha \, dA}{r^3} \tag{3.1.21}$$

In the numerator, we now find the scalar product of vectors $\vec{r}$ and $d\vec{A}$:

$$\Omega = \int_A \frac{\vec{r} \, d\vec{A}}{r^3} \tag{3.1.22}$$

or

$$\Omega = \int_A \frac{\vec{e}_N \, d\vec{A}}{r^2} \tag{3.1.23}$$

with $\vec{e}_N$ as unit vector in direction $\vec{r}$.

It applies for any origin of coordinates (Figure 3.1.5) that

$$\Omega = \int_A \frac{(\vec{r} - \vec{r}_0) d\vec{A}}{|\vec{r} - \vec{r}_0|^3} \tag{3.1.24}$$

The range of Ω lies between 4π and -4π. The solid angle is negative when the surface normal points towards space point P.

For smaller areas A that are perpendicular to the space axis a simple approximation is often sufficient:

$$\Omega \approx \frac{A}{r^2} \tag{3.1.25}$$

3.1.2.3 Circular Area with the Centre in The Distance Vector

The circular area has radius R_0 and distance r_0 to space point P (Figure 3.1.6). Thus, it follows for the solid angle that

$$\Omega = \int_A \frac{\cos \alpha}{R^2} \, dA \tag{3.1.26}$$

with

$$\cos \alpha = \frac{r_0}{R} \tag{3.1.27}$$

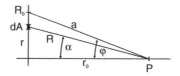

Figure 3.1.6   Circular area with the centre in the distance vector

$$\cos \varphi = \frac{r_0}{a} \tag{3.1.28}$$

$$R^2 = r^2 + r_0^2 \tag{3.1.29}$$

$$a^2 = R_0^2 + r_0^2 \tag{3.1.30}$$

and

$$dA = 2\,\pi\,r\,dr \quad \text{with} \quad 0 \le r \le R_0 \tag{3.1.31}$$

The solid angle is

$$\Omega = 2\pi \int\limits_0^{R_0} \frac{r_0}{R^3} r\,dr \tag{3.1.32}$$

$$\Omega = 2\,\pi\,r_0 \int\limits_0^{R_0} \frac{r}{(r^2+r_0^2)^{\frac{3}{2}}}\,dr = -2\,\pi\,r_0 \frac{1}{(r^2+r_0^2)^{\frac{1}{2}}}\Bigg|_0^{R_0} \tag{3.1.33}$$

$$\Omega = 2\pi \left(1 - \frac{r_0}{(R_0^2+r_0^2)^{\frac{1}{2}}}\right) = 2\pi \left(1 - \frac{r_0}{a}\right) \tag{3.1.34}$$

$$= 2\pi(1-\cos\varphi) = 4\pi \sin^2 \frac{\varphi}{2}$$

The result corresponds to that of the right circular cone (Equation 3.1.9). After projecting it to a sphere, the assumed circular area is the already calculated sphere cap. The calculation of the projection of an area to a unit sphere is often easier. An example would be a triangle situated in space.

3.1.2.4 Triangular Area Situated Arbitrarily in Space

The projection of a triangle arbitrarily situated in space to a unit sphere always results in a spherical triangle on the spherical surface (Figure 3.1.7).

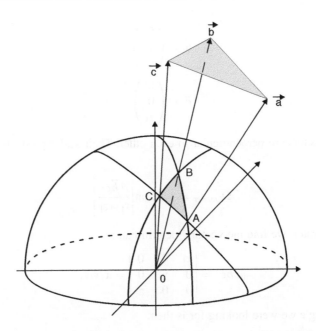

Figure 3.1.7   Projection of a plane triangle to the unit sphere

For a spherical radius of 1, the spherical triangle's area is equal to the solid angle we are looking for:

$$\Omega_{DE} = \alpha + \beta + \gamma - \pi \qquad (3.1.35)$$

with angles α, β and γ corresponding to the angles at points A, B and C in the spherical triangle. These angles can be calculated using vectors $\vec{a}, \vec{b}$ and $\vec{c}$ that form the triangle [1]. It results in the following simple relationship for solid angle Ω_{DE}:

$$\Omega_{DE} = 2\arctan \frac{|\vec{a}\vec{b}\vec{c}|}{abc + \left(\vec{a}\,\vec{b}\right)c + (\vec{a}\,\vec{c})b + \left(\vec{b}\,\vec{c}\right)a} \qquad (3.1.36)$$

The high importance of this triangle calculation is based on the fact that Equation 3.1.36 can be easily calculated numerically and that all areas can be approximated by the sum of triangles.

Example 3.1:   Solid Angle of a Triangular Area

The relationship in Equation 3.1.36 can be easily checked with a simple example. If the vectors that form the triangle are situated on the coordinate axes they cut out on the sphere surface a fourth of the half sphere. Solid angle $\Omega_{Quarter}$ thus is $1/2 \, \pi \, \Omega_0$. It is calculated as follows using Equation 3.1.36:

$$\vec{a} = \begin{pmatrix} x_1 \\ 0 \\ 0 \end{pmatrix} \qquad (3.1.37)$$

$$\vec{b} = \begin{pmatrix} 0 \\ y_1 \\ 0 \end{pmatrix} \qquad\qquad (3.1.38)$$

$$\vec{c} = \begin{pmatrix} 0 \\ 0 \\ z_1 \end{pmatrix} \qquad\qquad (3.1.39)$$

The three vectors are perpendicular to each other. Their scalar products are therefore zero:

$$\Omega_{\text{Quarter}} = 2\arctan\left|\frac{|\vec{a}\,\vec{b}\,\vec{c}|}{x_1 y_1 z_1}\right| \qquad (3.1.40)$$

In the numerator, we find the vectors' determinant:

$$|\vec{a}\,\vec{b}\,\vec{c}| = \begin{vmatrix} x_1 & 0 & 0 \\ 0 & y_1 & 0 \\ 0 & 0 & z_1 \end{vmatrix} = x_1 y_1 z_1 \qquad (3.1.41)$$

The solid angle we were looking for is then:

$$\Omega_{\text{Quarter}} = 2\arctan(1) = \frac{\pi}{2}\Omega_0 \qquad (3.1.42)$$

3.1.2.5 Rectangular Area

Rectangular area A is situated in the xy-plane (Figure 3.1.8). The rectangle has edge lengths X_0 and Y_0. Point P is situated on the z-axis. Angles φ_{X0} and φ_{Y0} describe the angle of point P in relation to X_0 and Y_0. Appendix C provides the detailed calculation of the solid angle.

The solid angle is:

$$\Omega_{\text{RE}} = \arcsin\frac{X_0\, Y_0}{A_X\, A_Y} \qquad (3.1.43)$$

with

$$A_X^2 = z_0^2 + X_0^2 \qquad (3.1.44)$$

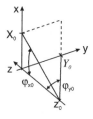

Figure 3.1.8   Position of the rectangular area

and

$$A_Y^2 = z_0^2 + Y_0^2 \qquad (3.1.45)$$

or

$$\Omega_{RE} = \arcsin\left[\sin(\varphi_{x0})\sin(\varphi_{y0})\right] \qquad (3.1.46)$$

3.2 Basic Law of Photometry

3.2.1 Definition

For two area elements dA_1 and dA_2 that are situated opposite each other (Figure 3.2.1a), radiant flux Φ_{21} that is emitted by area element dA_2 (emitter) and passes through area element dA_1 (receiver) is proportional to differential solid angle $d\Omega_2$:

$$d^2\Phi_{21} = L_2 \, dA_2 \, d\Omega_2 \qquad (3.2.1)$$

If both areas are situated parallel to each other, the differential solid angle is provided by Equation 3.1.15:

$$d^2\Phi_{21} = 2\,\pi\,\Omega_0 L_2 \, dA_2 \sin\varphi \, d\varphi \qquad (3.2.2)$$

Rotating or tilting the area element dA_1 with constant solid angle $d\Omega_1$ by angle β_1, we arrive at one emitter-side and one receiver-side interpretation option (see Figure 3.2.1b and c). Looking from receiver area dA_1 to an emitter, perceived area A_2 becomes larger with increasing angle β_1 (Figure 3.2.1b). For the now perceived area A_2' that is parallel to dA_1 at the same distance, it applies that:

$$A_2' = A_2 \cos\beta_1 \qquad (3.2.3)$$

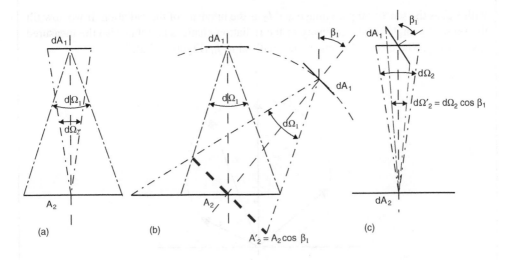

Figure 3.2.1 Geometric arrangement for the LAMBERT'S cosine law. (a) Receiver-side arrangement. (b) Receiver-side arrangement rotated on the unit sphere. (c) Emitter-side arrangement

Substituting the above relationship in Equation 3.2.1, the result is

$$d^2\Phi_{21} = L_2 \, dA_2 \cos \beta_1 \, d\Omega_2 \tag{3.2.4}$$

For a rotation of dA_1, an emitter (sender) perceives a smaller and smaller receiver area which results in a smaller solid angle Ω_2 for the receiver (Figure 3.2.1c):

$$\Omega_2' = \Omega_2 \cos \beta_1 \tag{3.2.5}$$

Substituting Equation 3.2.5 into Equation 3.2.1, we again arrive at Equation 3.2.4. For intensity I, it applies that:

$$I = \frac{d\Phi_{21}}{d\Omega_2} = L_2 \, dA_2 \cos \beta_1 \tag{3.2.6}$$

This relationship is also known as LAMBERT'S cosine law. LAMBERT'S cosine law states that an area with a constant radiance appears to have the same 'brightness' from all angles. When viewing from a receiver at an angle from above, you perceive more area elements of the emitter; they have an intensity that is reduced by the cosine of the viewing angle, though. Surfaces with a constant radiance L and with an irradiation that behaves according to the cosine law, are called LAMBERTian surfaces or LAMBERTian radiators.

Example 3.2:   Angle-Related Responsivity of Infrared Sensors

For measuring the responsivity of infrared sensors, they are arranged in front of a blackbody. If we know the emitter temperature and the distance between the sensitive area of the sensor and the emitter, we can determine the intensity of the radiation. An infrared sensor has a field of view (FOV) that is determined by its design. If sensor and emitter surfaces are situated parallel to each other, the following voltage responsivity R_V results (Section 5.1):

$$R_V = \frac{V_{out}}{I_0 \Omega_{FOV}} \tag{3.2.7}$$

with V_{out} as the sensor output voltage and I_0 as the intensity of the radiation. If we now tilt the sensor by angle β_1, the intensity of the radiation changes and thus also the measured responsivity. The changed intensity, normalised to I_0 (Figure 3.2.2), is:

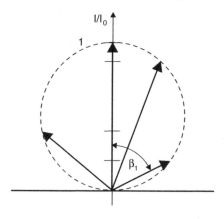

Figure 3.2.2   Changes in intensity due to tilting by angle β_1

$$\frac{I}{I_0} = \cos \beta_1 \qquad\qquad (3.2.8)$$

The measured responsivity decreases with increasing angle β_1.

Example 3.3:   Ideal Diffuse Reflection

Another example is the LAMBERTian surface that realises an ideal diffuse reflection. This means that the maximum of the reflected radiation is always found towards the surface normal, independent of the angle of incidence of the radiation. The intensity of the reflected radiation again depends on the angle's cosine in relation to the surface normal (Figure 3.2.3).

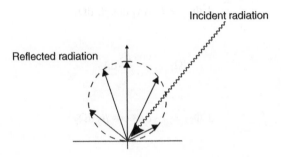

Figure 3.2.3   Ideal diffuse reflection on a LAMBERTian surface

---

If we now allow area element dA_2 to rotate, Equations 3.1.17 and 3.2.4 result in the basic law of photometry:

$$d^2\Phi_{12} = L_1 \frac{dA_1 \cos \beta_1 \, dA_2 \cos \beta_2}{r^2} \qquad\qquad (3.2.9)$$

or

$$d^2\Phi_{21} = L_2 \frac{dA_1 \cos \beta_1 \, dA_2 \cos \beta_2}{r^2} \qquad\qquad (3.2.10)$$

where r is the shortest distance between the two area elements (Figure 3.2.4). The basic law of photometry describes differential radiant flux $d\Phi_{12}$ that arrives from one area element dA_1 to area element dA_2 or differential radiant flux $d\Phi_{21}$, respectively, that moves from area element dA_2 to area element dA_1.

The radiant flux exchanged between the areas is

$$d^2\Phi_1 = d^2\Phi_{12} - d^2\Phi_{21} = \frac{dA_1 \cos \beta_1 \, dA_2 \cos \beta_2}{r^2} (L_1 - L_2) \qquad\qquad (3.2.11)$$

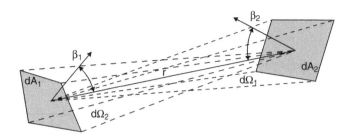

Figure 3.2.4 Definition of the geometric variables. dA_1, dA_2 area elements; r distance between area elements; β_1, β_2 angles between area elements and distance vector

Applying the differential solid angles, the Equations 3.2.9 and 3.2.10 can be expressed as follows:

$$d^2\Phi_{12} = L_1\, dA_1 \cos \beta_1\, d\Omega_1 \qquad (3.2.12)$$

with

$$d\Omega_1 = \frac{\cos \beta_2}{r^2}\, dA_2 \qquad (3.2.13)$$

and

$$d^2\Phi_{21} = L_2\, dA_2 \cos \beta_2\, d\Omega_2 \qquad (3.2.14)$$

with

$$d\Omega_2 = \frac{\cos \beta_1}{r^2}\, dA_1 \qquad (3.2.15)$$

The equations for the basic law of photometry have the following basic structure:

$$d^2\Phi_{12} = L_1\, dA_1\, d\omega_1 \qquad (3.2.16)$$

with ω_1 as the projected solid angle, where ω_1 is also called effective or weighted solid angle:[2]

$$\omega_1 = \int_A \frac{\cos \beta_1 \cos \beta_2}{r^2}\, dA_2 \qquad (3.2.17)$$

or respectively

$$\omega_{12} = \frac{1}{A_1} \int_{A_1} \int_{A_2} \frac{\cos \beta_1 \cos \beta_2}{r^2}\, dA_2\, dA_1 \qquad (3.2.18)$$

or

$$\omega_1 = \int_\Omega \cos \beta_1\, d\Omega_1 \qquad (3.2.19)$$

---

[2] ω_1 denotes the projected solid angle of area element dA_1 in relation to area A_2. ω_{12} is the projected solid angle of area A_1 in relation to area A_2.

or respectively

$$\omega_{12} = \frac{1}{A_1} \int\limits_{A_1} \int\limits_{\Omega} \cos \beta_1 \, d\Omega_1 \, dA_1 \tag{3.2.20}$$

The projected solid angles ω_{12} and ω_{21} can be easily transformed into each other:

$$\omega_{12}A_1 = \omega_{21}A_2 \tag{3.2.21}$$

The exchanged radiant flux thus becomes:

$$\Phi_1 = \Phi_{12} - \Phi_{21} = (L_1 - L_2)\omega_{12}A_1 = (L_1 - L_2)\omega_{21}A_2 \tag{3.2.22}$$

with ω_{21} as the projected solid angle of area A_2 in relation to area A_1.

It is also common to state the form factor [2]:

$$F_{12} = \frac{\omega_{12}}{\pi} \tag{3.2.23}$$

The form factor is also called angle or exchange factor. It is useful when we do not look at the radiance of a surface, but on its temperature. Equation 3.2.22, for instance, becomes:

$$d\Phi_1 = d\Phi_{12} - d\Phi_{21} = \sigma(T_1^4 - T_2^4) \, F_{12} \, dA_1 = \sigma(T_1^4 - T_2^4) \, F_{21} \, dA_2 \tag{3.2.24}$$

Example 3.4: Radiance of a Blackbody

Exitance M_{BB} is defined as (Figure 3.2.5):

$$M_{BB} = \frac{d\Phi_{12}}{dA_1} = L_1 \, \omega_1 = L_1 \int\limits_{\Omega} \cos \beta_1 \, d\Omega_1 \tag{3.2.25}$$

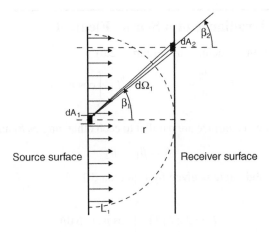

Figure 3.2.5 Geometric variables at a blackbody

As emitter and receiver surface are parallel to each other, angles β_1 and β_2 are the same:

$$\beta = \beta_1 = \beta_2 \tag{3.2.26}$$

The differential solid angle is given by Equation 3.1.15:

$$M_{BB} = 2\,\pi\,L_1\,\Omega_0 \int_0^\phi \cos\beta \sin\beta \, d\beta \tag{3.2.27}$$

Solving the integral, the exitance becomes

$$M_{BB} = \pi\,L_1\,\Omega_0 \sin^2\phi \tag{3.2.28}$$

The blackbody emits its radiation into the entire half-space

$$\varphi = \frac{\pi}{2} \tag{3.2.29}$$

Thus the exitance can be calculated as

$$M_{BB} = \pi\,L_1\,\Omega_0 \tag{3.2.30}$$

Applying the STEFAN–BOLTZMANN law

$$M_{BB} = \sigma T^4 \tag{3.2.31}$$

the radiance of a blackbody can now be calculated as a function of the temperature:

$$L_1 = \frac{\sigma}{\pi\,\Omega_0} T^4 \tag{3.2.32}$$

Example 3.5:   Irradiance of a Sensor Element

Irradiance E of area dA_2 is defined as (Figure 3.2.6)

$$E = \frac{d\Phi_{12}}{dA_2} = L_1\,\omega_2 = L_1 \int_\Omega \cos\beta_2 \, d\Omega_2 \tag{3.2.33}$$

As emitter and receiver surface are parallel to each other, angles β_1 and β_2 are the same:

$$\beta = \beta_1 = \beta_2 \tag{3.2.34}$$

The differential solid angle is given by Equation 3.1.15:

$$E = 2\,\pi\,L_1\,\Omega_0 \int_0^\phi \cos\beta \sin\beta \, d\beta \tag{3.2.35}$$

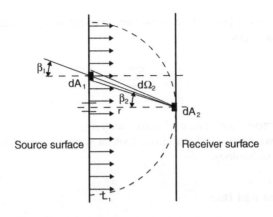

Figure 3.2.6   Geometric variables at a sensor element

and after solving the integral:

$$E = \pi L_1 \, \Omega_0 \sin^2 \varphi \tag{3.2.36}$$

or with the exitance:

$$E = M_S \sin^2\varphi = \sigma \, T^4 \sin^2 \varphi \tag{3.2.37}$$

It has to be taken into account that φ is half the field of view of the sensor element ($FOV/2$).

Comparing Examples 3.4 and 3.5, we can see that the calculations of radiance and irradiance arrive at the same result. For optical arrangements, in general the reversing principle applies. The projected solid angle can be calculated both from the receiver side and the emitter side.

3.2.2   Calculation Methods and Examples

3.2.2.1 Projected Solid Angle

The calculation of the projected solid angle is far more complicated than the calculation of the solid angle as the equation includes an additional trigonometric function ($\cos \beta_1$). In addition to a direct solution of Equations 3.2.9 or 2.3.10, we can also use STOKES' integral theorem to convert the surface integrals from Equation 3.2.9 into contour integrals [3]:

$$\omega_{12} = \frac{1}{2\,A_1} \oint_{C_2} \oint_{C_1} (\ln r \, dx_1 \, dx_2 + \ln r \, dy_1 \, dy_2 + \ln r \, dz_1 \, dz_2) \tag{3.2.38}$$

with C_1, C_2 as the boundary curves of areas 1 or 2, respectively, and r as the distance of the boundary curves. In [2], there are stated a number of calculation methods and form factors. [3.3] describes different numerical procedures. Complex areas can be calculated as the sum of non-overlapping individual areas:

$$\omega_{12} = \sum_n \omega_{12,n} \qquad (3.2.39)$$

with $\omega_{12,n}$ as the projected solid angle of partial area n.

In the following, we want to take a closer look at some projected solid angles that are important for sensor technology.

3.2.2.2 Area Element and Disc

Many calculations in sensor technology, such as calculation for the noise-equivalent temperature difference *NETD* (see section 5.4), are based on the simple model presented in Figure 3.2.7. It represents a small sensor element dA_1 that is illuminated by a black surface (blackbody or the entrance pupil of an optical device).

Projected solid angle ω_1 is according to [2]:

$$\omega_1 = \frac{\pi}{2} \left[1 - \frac{a^2 + h^2 - r^2}{\sqrt{(a^2 + h^2 + r^2)^2 - 4r^2 a^2}} \right] \qquad (3.2.40)$$

The case where the sensor element is situated in the centre of the disc $(a = 0)$ is of particular interest:

$$\omega_1 = \pi \frac{r^2}{h^2 + r^2} = \pi \sin^2 \varphi \qquad (3.2.41)$$

This important relationship was already calculated in Section 3.2.3. It is shown in Figure 3.2.8. If the area element sees the entire half space ($\varphi = 90°$), the projected solid angle becomes $\omega_1 = \pi$. For real arrangements, such as lenses, often f-number F is given:

$$F = \frac{h}{2r} \qquad (3.2.42)$$

Thus, Equation 3.2.41 becomes:

$$\omega_1 = \frac{\pi}{4F^2 + 1} \qquad (3.2.43)$$

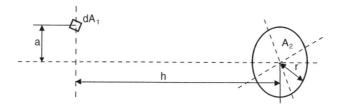

Figure 3.2.7   Arrangement with an area element dA_1 and a disc

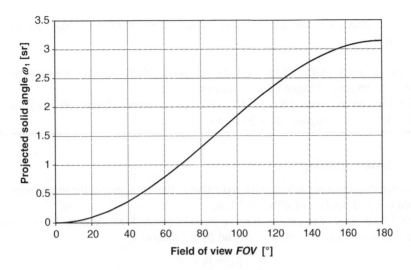

Figure 3.2.8 Projected solid angle of an area element

Example 3.6:   Projected Solid Angle of the Pixel of a Bolometer Array

In a sensor array, all pixels – with the exception of the central pixel – are situated outside the
optical axis. Figure 3.2.9 shows the projected solid angle in relation to distance a to the array

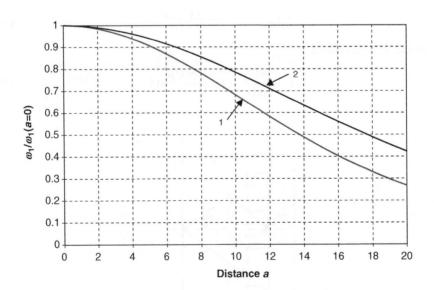

Figure 3.2.9 Projected solid angle of an area element in relation to the distance to the array centre,
normalised to $\omega_1(a=0)$. Curve 1: $r=9\,\text{mm}$; $h=18\,\text{mm}$ $(F=1)$. Curve 2: $r=9\,\text{mm}$; $h=25\,\text{mm}$
$(F=1.4)$

centre. The maximum deviation occurs in the diagonal of the array, for example (Table 6.6.8):

- Bolometer array with 384×288 pixels, pixel pitch $35\,\mu m$: Rectangular area $13.44 \times 10.08\,mm^2$; Diagonal $2a = 16.8\,mm$;
- Bolometer array with 640×480 pixels, pixel pitch $25\,\mu m$: Rectangular area $16.0 \times 12.0\,mm^2$; Diagonal $2a = 20.0\,mm$.

3.2.2.3 Natural Vignetting

In optics, natural vignetting refers to the reduction of the irradiance at the image edges. In the photometry, it is describes with the $\cos^4$ law.

Figure 3.2.10 shows the arrangement to be calculated. Area element dA_2 is situated at distance r' in relation to dA_1. Angles β_1 and β_2 are identical:

$$\beta_1 = \beta_2 = \beta \tag{3.2.44}$$

The basic law of photometry is then

$$d^2\Phi_{12} = L_1 \frac{dA_1\,dA_2\,\cos^2\beta}{r'^2} \tag{3.2.45}$$

with

$$r = \frac{r'}{\cos\beta} \tag{3.2.46}$$

it follows from the above equation that

$$d^2\Phi_{12} = L_1 \frac{dA_1\,dA_2\,\cos^4\beta}{r^2} \tag{3.2.47}$$

Irradiance E is calculated as

$$E = \frac{d\Phi_{12}}{dA_2} = L_1 \int_{A_1} \frac{\cos^4\beta}{r^2}\,dA_1 \tag{3.2.48}$$

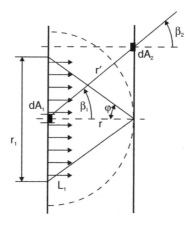

Figure 3.2.10 Geometric arrangement for calculating natural vignetting

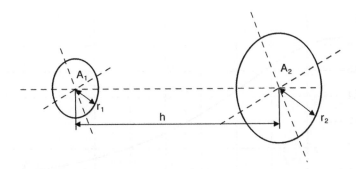

Figure 3.2.11   Model of two parallel disks on one and the same axis

Several approximations are used to arrive at the known $\cos^4$ law. Area A_2 is assumed to be circular (radius r_1) and independent of β:

$$A_1 = \pi\, r_1^2 \tag{3.2.49}$$

Thus, it results that

$$E = \pi\, L_1\, \Omega_0\, \frac{r_1^2}{r^2}\, \cos^4\beta \approx \pi\, L_1\, \Omega_0\, \sin^2\varphi\, \cos^4\beta \tag{3.2.50}$$

where φ is the half field of view of area element dA_2 at $\beta = 0$. The $\cos^4$ law only applies to small fields of view φ. Applying Equation 3.2.40 to the previous example results in the exact formula.

3.2.2.4 Two Parallel Discs

Figure 3.2.11 shows the arrangement of two parallel discs with their centres being situated on one axis. According to [2], the projected solid angle is:

$$\omega_{12} = \frac{\pi}{2}\left[X - \sqrt{X^2 - 4\left(\frac{r_2}{r_1}\right)^2}\right] \tag{3.5.51}$$

with

$$X = 1 + \frac{h^2 + r_2^2}{r_1^2} \tag{3.2.52}$$

Figure 3.2.12 shows an example. Another example is provided in Section 3.2.3.

3.2.2.5 Area Element and Rectangular Area

Figure 3.2.13 shows an arrangement of an area element dA_1 and a rectangular area. One corner of the rectangular area is situated on a line with area element dA_1 (distance h). The projected solid angle is according to [4]:

$$\omega_1 = \frac{1}{2}\left[\sin\varphi_k \cdot \arctan\left(\frac{m}{k}\sin\varphi_k\right) + \sin\varphi_m \cdot \arctan\left(\frac{k}{m}\sin\varphi_m\right)\right] \tag{3.2.53}$$

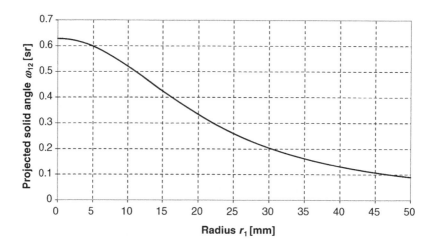

Figure 3.2.12 Projected solid angle of two parallel discs $r_2 = 9\,\mathrm{mm}$; $h = 18\,\mathrm{mm}$ (f-number $F = 1$); for $r_1 = 0$, it applies that $\omega_{12} = \omega_1$

with

$$\sin \varphi_k = \frac{k}{\sqrt{k^2 + h^2}} \tag{3.2.54}$$

and

$$\sin \varphi_m = \frac{m}{\sqrt{m^2 + h^2}} \tag{3.2.55}$$

Example 3.7: Projected Solid Angle of a Square and a Circular Aperture Stop

We will look at the difference between the projected solid angle of a rectangular and a circular aperture with the same area. The projected solid angle of a round aperture is given by Equation 3.2.41. For a square aperture with area element dA_1 in the optical axis, the projected solid angle has to be composed of four identical squares according to Figure 3.2.13. With

$$a = m = k \tag{3.2.56}$$

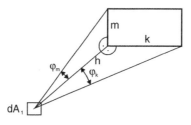

Figure 3.2.13 Arrangement of area element dA_1 and a rectangular area

and

$$\sin \varphi_a = \frac{a}{\sqrt{a^2 + h^2}} \qquad (3.2.57)$$

Equation 3.2.35 becomes

$$\omega_{1,\text{Quadrat}} = 4 \sin \varphi_a \arctan (\sin \varphi_a) \qquad (3.2.58)$$

Rectangular and circular area shall be of the same size:

$$\pi r^2 = 4a^2 \qquad (3.2.59)$$

With f-number F from Equation 3.2.42, it then results that:

$$\sin \varphi_a = \frac{1}{\sqrt{\dfrac{16}{\pi} F^2 + 1}} \qquad (3.2.60)$$

Figure 3.2.14 presents both projected solid angles. They lie exactly one on the other. The projected solid angle does not depend on the form of the area, but only on the surface area that can be perceived from another area or another area element (similar to the solid angle).

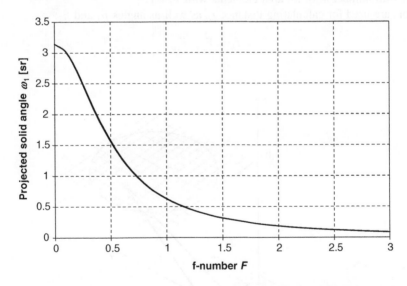

Figure 3.2.14 Projected solid angle of an area element pertaining to a rectangular or circular area (both curves lie one on the other)

3.2.3 Numerical Solution of the Projected Solid Angle

Already the calculation of the projected solid angle of a simple arrangement, such as a rectangular area in relation to a displaced rectangular area, results in complicated integrals that cannot be solved using analytical methods. In sensorics, we often have to calculate defined circular or rectangular arrangements of sensor and emitter area or entrance pupil of an optical device. These areas do not touch, intersect or cover each other and are self-contained. Here, the principle of finite area elements can be used as a simple numerical solution for such an arrangement [3].

At first, we will look at the projected solid angle of differential area element $dA_{1,i}$ in relation to area A_2, that can be decomposed into n finite area elements $\Delta A_{2,j}$ (Figure 3.2.15):

$$\omega_{1,i} = \sum_{j=1}^{n} V_j \frac{\cos \beta_{1,ij} \cos \beta_{2,ij}}{r_{ij}^2} \Delta A_{2,j} \qquad (3.2.61)$$

Factor V_j states whether area element $\Delta A_{2,j}$ belongs to area A_2 ($V_j = 1$) or not ($V_j = 0$). The advantage is that area A_2 can have any form and that the area can be decomposed into square finite area elements. Here it applies that:

Number of columns i = number of rows j,

$$n = i \cdot j \text{ and} \qquad (3.2.62)$$

$$\Delta A_{2,j} = \frac{A_2}{n'} \qquad (3.2.63)$$

for all j, with number n' of all area elements with $V_j = 1$.

Vectors are used for calculating distance r_{ij} as well as angles β_1 and β_2:

$$\vec{r}_{ij} = \vec{r}_{2,j} - \vec{r}_{1,i} \qquad (3.2.64)$$

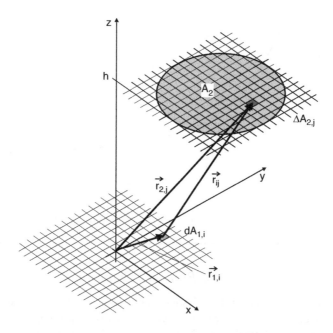

Figure 3.2.15   Geometric arrangement for the numerical solution of a projected solid angle

with $\vec{r}_{1,i}$ and $\vec{r}_{2,j}$ being the vectors of area elements $dA_{1,i}$ and $\Delta A_{2,j}$. In addition, it applies that

$$\cos \beta_{1,ij} = \frac{\vec{r}_{ij} \cdot \vec{e}_{A,1i}}{r_{ij}} \qquad (3.2.65)$$

and

$$\cos \beta_{2,ij} = \frac{-\vec{r}_{ij} \cdot \vec{e}_{A,2j}}{r_{ij}} \qquad (3.2.66)$$

with $\vec{e}_{A,1i}$ and $\vec{e}_{A,2j}$ as the surface normals of both area elements.

The calculations become less complex if we position areas $dA_{1,i}$ and A_2 in the x-y-plane ($dA_{1,i}$ for $z = 0$ and A_2 for $z = h$):

$$\vec{r}_{1,i} = \begin{pmatrix} x_{1,i} \\ y_{1,i} \\ 0 \end{pmatrix} \qquad (3.2.67)$$

$$\vec{r}_{2,j} = \begin{pmatrix} x_{2,j} \\ y_{2,j} \\ h \end{pmatrix} \qquad (3.2.68)$$

$$\vec{e}_{A,1i} = \vec{e}_z \qquad (3.2.69)$$

$$\vec{e}_{A,2j} = -\vec{e}_z \qquad (3.2.70)$$

Then, angles $\beta_{1i,k}$ and $\beta_{2j,k}$ also have the same value (alternate angles on parallels), and it applies that:

$$\omega_{1,i} = \frac{A_2}{n'} \sum_{j=1}^{n} V_j \frac{h^2}{\left[(x_{2,j} - x_{1,i})^2 + (y_{2,j} - y_{1,i})^2 + h^2 \right]^2} \qquad (3.2.71)$$

or

$$\omega_{1,i} = \frac{A_2}{h^2} \frac{1}{n'} \sum_{j=1}^{n} V_j \frac{1}{\left(\frac{r_{ij}^2}{h^2} + 1 \right)^2} \qquad (3.2.72)$$

Figure 3.2.16 present the projected solid angle of area element dA_1 of circular area A_2 in relation to the number of area elements n. For a comparison, we also provide the exact value from Equation 3.2.43. For value $n = 1$, we used the approximate value from Equation 3.1.25. MATLAB was used for the computation. The required number of area elements is low. Using current computation technology, for 10 000 elements calculation time was less than 1 s.

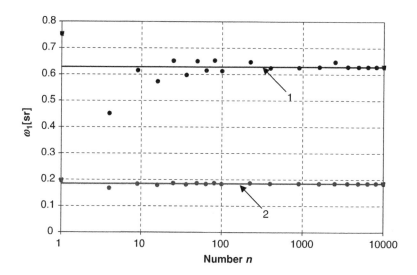

Figure 3.2.16 Projected solid angle of an area element dA_1 in relation to circular area A_2 solid line: exact calculation; dots: numerical solution $x_{1,i} = y_{1,i} = 0$; Curve 1: $h = 18$ mm, $r_2 = 9$ mm (f-number $F = 1$); Curve 2: $h = 36$ mm, $r_2 = 9$ mm (f-number $F = 2$)

The projected solid angle of area A_1 in relation to area A_2 can now be calculated decomposing area A_1 into m finite area elements, too, and using the above formula for each area element:

$$\omega_{12} = \frac{1}{A_1} \sum_{i=1}^{m} \sum_{j=1}^{n} W_i \, V_j \frac{\cos \beta_{1,ij} \cos \beta_{2,ij}}{r_{ij}^2} \Delta A_{1,i} \Delta A_{2,j} \qquad (3.2.73)$$

Factor W_i states whether area element $\Delta A_{1,i}$ belongs to area element A_1 ($W_i = 1$) or not ($W_i = 0$). Given the above conditions (parallel areas) and

$$A_1 = m' \Delta A_{1,i} \qquad (3.2.74)$$

with m' as the number of all area elements and $W_i = 1$, ω_{12} becomes:

$$\omega_{12} = \frac{A_2}{m' n'} \sum_{i=1}^{m} \sum_{j=1}^{n} W_i \, V_j \frac{h^2}{[(x_{2,j} - x_{1,i})^2 + (y_{2,j} - y_{1,i})^2 + h^2]^2} \qquad (3.2.75)$$

Projected solid angle ω_{12} of area A_1 in relation to area A_2 is also the mean value of projected solid angle $\omega_{1,i}$ of all area elements $dA_{1,i}$ in relation to area A_2:

$$\omega_{12} = \frac{1}{m'} \sum_{i=1}^{m} W_i \omega_{1,i} \qquad (3.2.76)$$

Example 3.8:   Projected Solid Angle of Two Parallel Circular Areas

Two circular areas with radii $r_1 = r_2 = 1$ mm are situated opposed to each other (see also Figure 3.2.11). They are situated at distance $h = 1$ mm of each other. We use the procedure just described in Equation 3.2.73 to calculate projected solid angle ω_{12} in relation to distance x of the centre of the areas. Figure 3.2.17 shows the result. For comparison, we have also provided the projected solid angle ω_1 that was calculated according to Section 3.2.2.2. Here, area A_1 was assumed to be the area element. For $x = 0$, the calculation of projected solid angle ω_{12} according to Section 3.2.2.4 (parallel discs) results in

$$\omega_{12} = 1.20\,\Omega_0. \qquad (3.2.77)$$

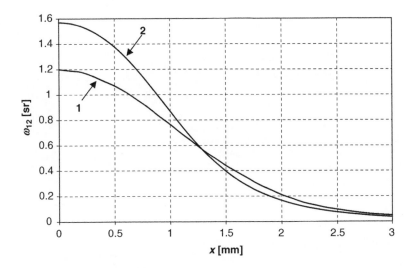

Figure 3.2.17   Projected solid angle ω_{12} of two parallel circular disks $r_1 = r_2 = 1$ mm; $h = 1$ mm; displaced by x; Curve 1: Numerical solution; Curve 2: Approximation according to Section 3.2.2.2 (area element dA_1 in relation to the circular disk)

References

1. van Oosterom, A. and Strackee, J. (1983) The solid angle of a plane triangle. *IEEE Transactions on Biomedical Engineering*, **BME-30**(2), 125–126.
2. Siegel, R., Howell, J.R. and Lohrengel, J. (1991) Wärmeübertragung durch Strahlung, in *Teil 2: Strahlungsaustausch zwischen Oberflächen und in Umhüllungen (Heat Transfer by Radiation: Radiation Exchange between Surfaces and in Covers)*, Springer-Verlag, Berlin.
3. Cohen, M.F. and Wallace, J.R. (1993) *Radiosity and Realistic Image Synthesis*, Morgan Kaufmann Publishers, San Francisco, California.
4. Vincent, J.D. (1989) *Fundamentals of Infrared Detector Operation and Testing*, John Wiley & Sons Ltd, Chichester.

4

Noise

The measuring uncertainty of infrared sensors und subsequently their resolution regarding the measured parameters (e.g. radiant flux, temperature, amongst others) is basically determined by noise processes. In the following, we will therefore present mathematical and physical basics of noise processes in thermal infrared sensors.

4.1 Mathematical Basics

4.1.1 Introduction

When we talk about noise we actually mean small random variations of a signal.[1] It has an unregular, random time pattern that cannot be predicted. Noise is a stochastic process. Mathematically, distribution and density functions can be used to describe stochastic processes. Noise can therefore be described with expected value E and variance σ^2 or distribution σ. Inserting measured noise values in a histogram, we arrive at a frequency distribution (Figures 4.1.1 and 4.1.2). We distinguish between continuous (steady) and discrete noise processes.

Continuous processes can adopt any amplitude value. An example would be thermal noise (Figure 4.1.1). Continuous processes often follow a normal distribution.

Discrete processes can adopt only specific values that are integer multiples of a physical parameter. An example would be current noise. In this case, there occur fluctuations, that is, the random creation and extinction of charge. The frequency of this process follows a POISSON distribution (Figure 4.1.2).

When using an analogue-to-digital converter (ADC) for measuring a noise voltage, we arrive at time- and amplitude-discrete values (digital values). If we choose the correct resolution of ADC, the measurement-caused discretisation of the amplitudes can be neglected. It should be noted that intermediate values are unknown for the discrete time interval of the measurement, that is, the amplitude response between two measuring points is undefined. Consequently,

---

[1] "Signal" refers to an electric charge, an electric current, a photon flux, etc. The short term "noise" usually describes a noise voltage or a noise current.

Thermal Infrared Sensors: Theory, Optimisation and Practice   Helmut Budzier and Gerald Gerlach
© 2011 John Wiley & Sons, Ltd

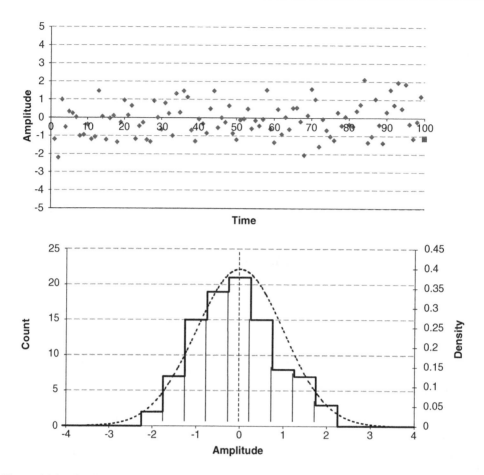

Figure 4.1.1 Continuous noise (normal distribution). (a) 100 normally distributed measured values of the amplitude of a measured parameter; (b) histogram of the measuring values from (a) (solid line) and theoretical density (broken line) with expected value $E = 0$ and variance $\sigma = 1$ (normalised normal distribution)

the measuring points cannot be connected. Figures 4.1.1 and 4.1.2 therefore show the measured values as dots and not as a curve. The normal distribution is a continuous distribution; thus, Figure 4.1.1 represents density as a closed curve. Any amplitude value is possible even if it has not been measured. For a discrete POISSON distribution, there are only specific amplitudes. Frequency is only marked for these values (Figure 4.1.2). There are no intermediate values. If we use digital values for calculating the continuous noise, we apply the methods of discrete distribution. The results are then interpreted as a continuous phenomenon. Discrete distributions can be considered a special case of continuous distribution.

For all further considerations, we assume two important conditions for processes or signals, respectively:

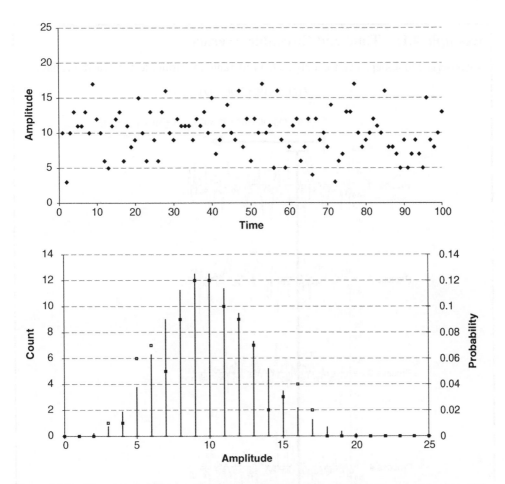

Figure 4.1.2 Discrete noise (POISSON distribution). (a) 100 POISSON-distributed measured values of the amplitude of a measured parameter with discrete values; (b) histogram of the measured values from (a) (squares) and theoretical density function (lines) with expected value $E = 10$

- **Stationarity:** A signal is stationary if the mean value of signal is independent at the time of measurement and its autocorrelation function only depends on displacement time τ. Stationarity can be easily checked by repeatedly measuring the mean value at any time. The results have to be identical. Examples of non-stationary signals are switch-on and transient processes and processes following the change of an operational parameter, such as temperature, operating point and so on.
- **Ergodicity:** A system is called ergodic if it is stationary and the time average corresponds to the average at a specific time over a large number of different process realisations, the so-called ensemble average. Checking ergodicity requires substantial equipment and time.

Example 4.1:   Time and Ensemble Average

We observe a time-dependant sine signal $f(t)$ with random amplitude A and random phase φ:

$$f(t) = A\sin\left(\omega t + \varphi\right) \tag{4.1.1}$$

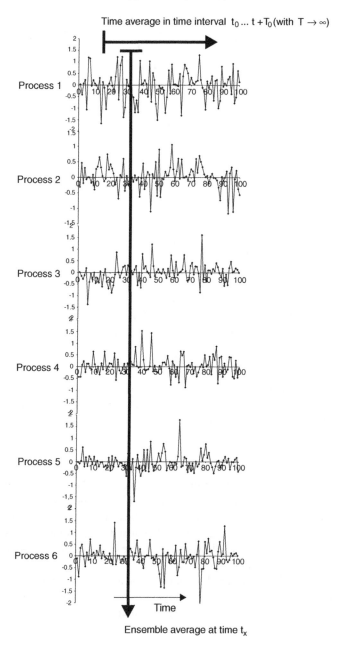

Figure 4.1.3   Time and ensemble average

Amplitude A and phase φ follow a distribution ($E=0$; $\sigma=1$). For the time average, it applies that

$$\overline{f(t)} = \lim_{T \to \infty} \frac{1}{2T} \int_{-T}^{T} f(t)\,dt = 0 \qquad (4.1.2)$$

or, respectively, for digital, that is, discretely sampled values:

$$\overline{f(t)} = \frac{1}{N} \sum_{n=1}^{N} f(t_n) = 0 \qquad (4.1.3)$$

Here, N has to be very large. It applies for the ensemble average that

$$\langle f_{t_x}(k) \rangle = \frac{1}{M} \sum_{m=1}^{M} f(k_m) = 0 \qquad (4.1.4)$$

with $f_{t_x}(k)$ as the measuring value of function f at time t_x and realisation k. Figure 4.1.3 illustrates the situation. The time average can be measured for any process starting at any selected time t_0. The ensemble average also is calculated at any time t_x by measuring M processes. Here, M has to be very large.

As the following example of noise voltage measurements at resistors shows, determining the ensemble average requires substantial equipment. For small-band measurements (bandwidth $= 1\,\text{Hz}$), we get a measured signal according to Equation 4.1.1, as presented in Figure 4.1.3. We can now measure at one resistor the time average according to Equation 4.1.3. For determining the ensemble average, we require M similar resistors, for example, $M = 1000$. Using M measuring devices, we will now simultaneously measure the amplitude at time t_x. Figure 4.1.3 shows six realisations. We can now use Equation 4.1.4 to calculate the ensemble average. Ensemble average and time average should be identical.

Finally, it should be noted that for testing stationarity and ergodicity, the above described test of the mean values is not sufficient. We also have to analyse instants of higher order, such as variance.

In addition to probability functions, we can also use time, correlation and spectral function to describe noise. In electrical engineering/electronics, spectral functions are particularly important for the description of noise, as it is the basis for the computation and simulation of electronic components and circuits. In the following sections, we will present all four possible signal descriptions.

Power is the central intersection of all mentioned options for the description of random signals. Power P that is converted in resistor R results with voltage V that is applied to the resistor and current I flowing through it in

$$P = VI = RI^2 = \frac{V^2}{R} \qquad\qquad (4.1.5)$$

In practice, we often only use the square value of the current I^2 or of the voltage V^2 to describe the power. The unit of the thus defined power is V^2 or A^2, respectively. In order to arrive at a power in physical terms (in Watt), we can normalise voltage and current square values to the resistance of $1\,\Omega$.

4.1.2   Time Functions

For a random signal, it is not possible to state a determined time function. An exact description of the time curve is only possible if it is measured afterwards. This is of fundamental importance as this is exactly what traditional measuring systems do (Figure 4.1.4).

An analogue-to-digital converter is used to read a sensor signal into the PC. The signal results from the superimposition of sensor signal $V_S(t)$ and random signal (sensor noise) $V_R(t)$:

$$V(t) = V_S(t) + V_R(t) \qquad\qquad (4.1.6)$$

Signal V_S is the value we are looking for whereas noise voltage V_R is a disturbance variable that does not contain any measuring information and therefore has to be eliminated. The PC then provides the digitised measuring signal:

$$V(t) = \sum_{n=1}^{N} [V_S(t_n) + V_R(t_n)] \qquad\qquad (4.1.7)$$

As we only want to look at the noise, we assume that signal $V_S(t)$ is determined and known. The noise can therefore be deducted from the measured value. The result is the noise voltage we are looking after:

$$V_R(t) = V(t) - V_S(t) \qquad\qquad (4.1.8)$$

In the following we will describe the parameters that are used for describing a stochastic signal. For general signal function f, we can insert voltage V or current I, respectively.

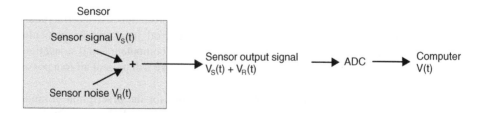

Figure 4.1.4   Typical sensor measuring system

4.1.2.1 Time Mean Value

The mean value describes the offset of a signal. It corresponds to the expected value:

$$\overline{f(t)} = \lim_{T \to \infty} \frac{1}{2T} \int_{-T}^{T} f(t)\, dt \tag{4.1.9}$$

or, respectively

$$\overline{f(t)} = \frac{1}{N} \sum_{n=1}^{N} f_n \tag{4.1.10}$$

Abbreviation f_n refers to function value f at the corresponding sampling time t_n:

$$f_n = f(t_n) \tag{4.1.11}$$

4.1.2.2 Square Time Mean Value

According to Equation 4.1.5, the square mean value corresponds to a signal's power:

$$\overline{f^2(t)} = \lim_{T \to \infty} \frac{1}{2T} \int_{-T}^{T} f^2(t)\, dt \tag{4.1.12}$$

or, respectively

$$\overline{f^2(t)} = \frac{1}{N} \sum_{n=1}^{N} f_n^2 \tag{4.1.13}$$

In electrical engineering, the square root of the square mean value, that is, the root mean square (rms) is of particular importance as this parameter can be used to compare noise signal and sensor signal:

$$\tilde{f} = f_{\text{rms}} = \sqrt{\overline{f^2(t)}} \tag{4.1.14}$$

4.1.2.3 Variance

Variance is a measure of the deviation from a mean value:

$$\sigma^2 = \overline{f^2(t)} - \overline{f(t)}^2 = \overline{\left(f(t) - \overline{f(t)} \right)^2} \tag{4.1.15}$$

or, respectively

$$\sigma^2 = \frac{1}{N} \sum_{n=1}^{N} (f_n - \overline{f(t)})^2 \tag{4.1.16}$$

As the difference has a quadratic behaviour, large deviations have a much larger effect on the variance than small deviations. The square root of the variance is the distribution. For a mean value of zero, the variance corresponds to the root mean square and thus to the power of the noise signal. For the measurement, the following formula for the calculation of the variance is suitable:

$$\sigma^2 = \frac{1}{N}\sum_{n=1}^{N} f_n^2 - \left(\frac{1}{N}\sum_{n=1}^{N} f_n\right)^2 \qquad (4.1.17)$$

The advantage of this formula consists in that it does not require saving all individual measured values, but only the current sum of the measured values and their square values.

4.1.3 Probability Functions

It is not possible to exactly predict function values of random signals. However, it is possible to state probability P of a function value occurring at a predetermined time within a specific interval $(x_0; x_1)$:

$$P(x_0 \leq x \leq x_1) = \int_{x_0}^{x_1} w(x)\,dx \qquad (4.1.18)$$

Probability density $w(x)$ describes the probability distribution of function values x. Here, the necessary condition applies that maximum probability $w(x)$ of all events occurring (with $x = -\infty \ldots + \infty$) amounts to 100% or unity, respectively:

$$\int_{-\infty}^{+\infty} w(x)\,dx = 1 \qquad (4.1.19)$$

The density can be used to calculate distribution function $W(x)$:

$$W(x) = P(x \leq x_0) = \int_{-\infty}^{x_0} w(x)\,dx \qquad (4.1.20)$$

For discrete random variables, the distribution function applies analogously:

$$W(x) = P(x \leq x_0) = \sum_{k=1}^{x_0} p_k \qquad (4.1.21)$$

given the condition

$$\sum_{k} p_k = 1 \qquad (4.1.22)$$

with individual probability

$$p_k = P(x_k) \qquad (4.1.23)$$

Expected value E and variance σ^2 are important characteristic parameters of the distribution and density functions:

$$E = \int_{-\infty}^{+\infty} x\,w(x)\mathrm{d}x \qquad (4.1.24)$$

or, respectively

$$E = \sum_{k=1}^{\infty} x_k p_k \qquad (4.1.25)$$

Variance σ^2 is also used to describe the magnitude of the variation around the expected value:

$$\sigma^2 = \int_{-\infty}^{+\infty} (x-E)^2 w(x)\,\mathrm{d}x \qquad (4.1.26)$$

or, respectively

$$\sigma^2 = \sum_{k=1}^{\infty} (x_k-E)^2 p_k \qquad (4.1.27)$$

The square root of the variance is distribution σ. In practice, expected value E and distribution σ is calculated based on the measured values, that is, they are estimated due to the existing measuring uncertainties. The corresponding estimated values for E and σ are mean value μ and standard deviation s.

For practical calculations, the following relations have proved useful:

$$E(ax+b) = aE(x)+b \qquad (4.1.28)$$

with constants a and b;

$$E(E_1+E_2) = E_1+E_2 \qquad (4.1.29)$$

$$E(E_1E_2) = E_1E_2 \qquad (4.1.30)$$

if E_1 and E_2 are expected values of two independent random functions as well as

$$\sigma^2(ax+b) = a^2\sigma^2(x) \qquad (4.1.31)$$

$$\sigma^2(x+y) = \sigma^2(x)+\sigma^2(y) \qquad (4.1.32)$$

if x and y are independent random variables.

We will use the following two examples to present two important distribution functions that are often related to noise processes.

4.1.3.1 Normal Distribution

The normal or GAUSSian distribution is the most well-known and probably the most frequent distribution function. This is due to the central limit theorem of the probability calculation:

Independent of the distribution of a random variable, the sum of n identically distributed, independent variables for $n \rightarrow \infty$ will be normally distributed.

As many noise mechanisms are based on the superimposition of several natural processes, they are often normally distributed. Two superimposed normally distributed random variables are again normally distributed.

The density function of the normal distribution is the famous GAUSSIAN bell curve:

$$w(x) = \frac{1}{\sigma\sqrt{2\pi}} e^{-\frac{(x-E)^2}{2\sigma^2}} \qquad (4.1.33)$$

The standard deviation σ is the distance of the inflection point of the bell curve from the expected value. Figure 4.1.5 presents the density and density distribution functions (see also the example in Figure 4.1.1). When expected value $E=0$ and variance $\sigma^2=1$, the normal distribution is called standard normal distribution.

The density function is symmetrical to the expected value. This allows statements regarding the statistical certainty of a function value being located within a certain range $\pm v$ around the expected value:

$$P(|x-E| \leq v) = 2\int\limits_{E}^{v} w(x)\,\mathrm{d}x \qquad (4.1.34)$$

Table 4.1.1 summarises some solutions of Equation 4.1.34. In addition it should be noted, though, that there can occur very large deviations from the expected value. Thus, the probability of a function value of $E \pm 10\sigma$ corresponds to the uncredibly small value 7.8×10^{-23}.

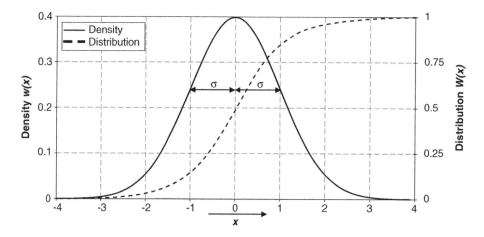

Figure 4.1.5   Distribution function $W(x)$ and density function $w(x)$ of the standard normal distribution ($E=0$ und $\sigma^2=1$)

Table 4.1.1   Statistical likelihood of function values

Range	Likelihood of the value lying within this range	Likelihood of the value lying outside this range
$E \pm 0.67\,\sigma$	0.500	0.500
$E \pm \sigma$	0.683	0.317
$E \pm 2\sigma$	0.953	0.047
$E \pm 3\sigma$	0.997	0.003

4.1.3.2 POISSON Distribution

For random experiments, where the results can only adopt specific values (e.g. 'yes' or 'no'; 'state 0 ', 'state 1', . . . 'state k'), there are discrete probability distributions. If – in addition – the probability of some of these discrete events is very small, the POISSON distribution provides a good approximation of the corresponding probability distribution. The POISSON distribution is therefore sometimes even called distribution of rare events.

Many discrete physical processes, such as the emission of light quanta, follow the POISSON distribution. If a radiation emission, for instance, produces one light quantum per millisecond, but the sensor that receives the radiation is read out only every 10 ms (i.e. the light quanta are only counted every 10 ms), we expect the following mean value per measurement

$$\lambda = 1 \text{ light quantum/ms} * 10 \text{ ms} = 10 \text{ light quanta}$$

Distribution function $W(k)$ or probability density function $P(k)$ of discrete state k is determined by parameter λ which is also called occurrence rate ($\lambda > 0$):

$$P(k) = p_k = \frac{\lambda^k}{k!} e^{-\lambda} \tag{4.1.35}$$

The distribution function of the probability of a state between state 0 and state k occurring is

$$W(k) = \sum_{k=0}^{N} p_k = \sum_{k=0}^{k} \frac{\lambda^k}{k!} e^{-\lambda} \tag{4.1.36}$$

The particularity of the POISSON distribution consists in that the expected value equals the variance:

$$E = \sigma^2 = \lambda \tag{4.1.37}$$

Figure 4.1.6 presents an example of a POISSON distribution (see also the example in Figure 4.1.2). Two superimposed POISSON-distributed random variables with parameters λ_1 and λ_2 are also POISSON-distributed – with parameter λ:

$$\lambda = \lambda_1 + \lambda_2 \tag{4.1.38}$$

For a large number n of independent occurrences with probability p, the POISSON distribution results from the binomial distribution:

$$\lambda = np \tag{4.1.39}$$

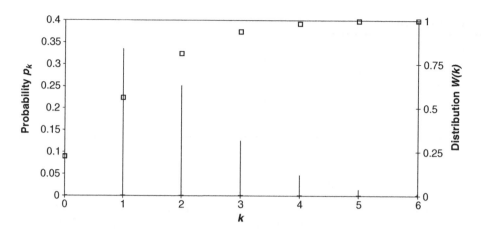

Figure 4.1.6 POISSON distribution for $\lambda = 1.5$. Squares: distribution $W(k)$; lines: individual probabilities p_k.

Example 4.2:   Noise of an Electric Current

Current noise is a POISSON distributed process (see Section 4.2.2). Mean current I_0 is

$$I_0 = \bar{N}\frac{q}{\tau} \tag{4.1.40}$$

with average number $\bar{N}$ of charge carriers q per time τ. Correspondingly, the instantaneous value of current I is

$$I = N\frac{q}{\tau} \tag{4.1.41}$$

with number N of charge carriers q per time τ. It applies for the variance that

$$\sigma^2 = \overline{(I-I_o)^2} = \frac{q^2}{\tau^2}\overline{(N-\bar{N})^2} \tag{4.1.42}$$

As for a POISSON distribution the expected value – that corresponds to the mean value – equals the variance, it follows that

$$\overline{(N-\bar{N})^2} = \bar{N} \tag{4.1.43}$$

The mean number of charge carriers can be calculated based on Equation 4.1.40:

$$\bar{N} = I_0\frac{\tau}{q} \tag{4.1.44}$$

and the variance results in

$$\sigma^2 = \frac{q}{\tau}I_0 \tag{4.1.45}$$

With equivalent noise bandwidth B_{eq} (see Equation 4.1.82):

$$B_{eq} = \frac{1}{2\tau} \qquad (4.1.46)$$

we arrive at the noise power of the current noise:

$$\bar{i}_{R,I}^2 = \sigma^2 = 2qI_0 B_{eq} \qquad (4.1.47)$$

Here the unit of the 'noise power' is A^2.

4.1.4 Correlation Functions

4.1.4.1 Autocorrelation Function

The autocorrelation function (*ACF*) states the internal relationship of a signal, that is, the correlation to itself. Looking at a signal at two points in time t_1 and t_2, that are offset by distance τ, the joint probability $w(x_1, x_2) = w(x, \tau)$ states the probability of x_1 at time t_1 equalling x_2 at time t_2:

$$ACF = \int_{-\infty}^{+\infty} \int_{-\infty}^{+\infty} x_1 x_2 w(x_1, x_2, \tau) \, dx_1 \, dx_2 \qquad (4.1.48)$$

Here, the probability density

$$w(x_1, x_2, \tau) = \lim_{\substack{\Delta x_1 \to 0 \\ \Delta x_2 \to 0}} \frac{P[x_1 < x < (x_1 + \Delta x_1); x_2 < x < (x_2 + \Delta x_2)]}{\Delta x_1 \Delta x_2} \qquad (4.1.49)$$

states probability P of the signal lying at time t_1 in the interval $(x_1 + \Delta x_1)$ and at time t_2 in the interval $(x_2 + \Delta x_2)$. Using time functions, *ACF* can be calculated as

$$ACF = \lim_{T \to \infty} \frac{1}{2T} \int_{-T}^{T} f(t) f(t + \tau) \, dt \qquad (4.1.50)$$

or for discrete sampling as

$$ACF(\tau) = \frac{1}{N} \sum_{n=1}^{N} f(n) f(n + \tau). \qquad (4.1.51)$$

ACF is an even function, that is, it applies that

$$ACF(\tau) = ACF(-\tau) \qquad (4.1.52)$$

The maximum of *ACF* lies always at $\tau = 0$, as there the function is mapped onto itself. As the comparison of Equations 4.1.12 and 4.1.50 shows, the value of *ACF* at point $\tau = 0$ is equal to the variance:

$$\sigma^2 = ACF(\tau = 0) \qquad (4.1.53)$$

The periodicity of $x(t)$ is strongly reflected in ACF. In addition, it generally applies that

$$ACF(\tau \rightarrow \infty) = 0 \qquad (4.1.54)$$

The autocorrelation function is a continuous function, also for discrete $f(t)$. The only exception occurs when ACF is a DIRAC pulse δ at time $\tau = 0$.

Example 4.3:   ACF of a Stochastic Signal

For purely stochastic signals without any internal relationship between the individual function values, ACF is

$$ACF(\tau) = \sigma^2 \delta(\tau) \qquad (4.1.55)$$

and thus

$$ACF(\tau \neq 0) = 0. \qquad (4.1.56)$$

Value $ACF = 0$ describes that it is not possible to derive at any time from a value at time τ another function value at any other time.

4.1.4.2 Cross Correlation

The cross correlation function (CCF) describes a statistical dependence of two signals $x(t)$ and $y(t)$:

$$CCF_{xy}(\tau) = \lim_{T \rightarrow \infty} \frac{1}{2T} \int_{-T}^{+T} x(t)\, y(t + \tau)\, dt \qquad (4.1.57)$$

It applies that

$$CCF_{xy} = CCF_{yx} \qquad (4.1.58)$$

Usually:

$$CCF_{xy}(\tau) \neq CCF_{xy}(-\tau) \qquad (4.1.59)$$

If

$$CCF(\tau) = 0$$

both variables are statistically independent.

There is a close relationship between cross correlation and covariance function. If the mean values of statistical signals x and y are zero, the covariance equals the cross correlation functions. If both random signals x and y are equal, the result is the autocorrelation function.

4.1.5 Spectral Functions

Usually, in electrical engineering and electronics we use the FOURIER representation of deterministic functions in order to describe signals. The corresponding statistical analogy is the spectral representation of stationary signals using spectral power density $S(f)$. This is the mean power per hertz bandwidth at frequency f that is converted in a load resistance of $1\,\Omega$. The total power then results from all power contributions of all frequency ranges between $-\infty \le f \le +\infty$:

$$P = \int_{-\infty}^{+\infty} S(f)\, df = \frac{1}{2\pi} \int_{-\infty}^{+\infty} S(\omega)\, d\omega \qquad (4.1.60)$$

The negative frequency range in this equation results from the general mathematically orientated representation, that always assumes function ranges of $-\infty \ldots +\infty$ (a so-called two-sided spectrum). However, only positive frequencies $0 \ldots +\infty$ (one-sided spectrum) are physically reasonable. Many theoretical calculations, such as the FOURIER transform, require a two-sided spectrum.

We have to assume that power P of a signal is finite even over an infinite time interval. In practice, this has to apply as the used frequency band has an upper limit due to sampling time t_A ($f_O = 1/(2t_A)$) and a lower limit due to the length of the observation interval or measuring time t_B ($f_u = 1/t_B$). The square root of the 'power'" in Equation 4.1.60 (voltage or current square value) is the root mean square:

$$\tilde{v} = \sqrt{P} \qquad (4.1.61a)$$

or, respectively

$$\tilde{i} = \sqrt{P} \qquad (4.1.61b)$$

The spectral power density $S(f)$ has the units $\dfrac{V^2}{Hz}$ or $\dfrac{A^2}{Hz}$, respectively. Often we do not use directly the spectral power density, but the square root of the spectral power density, the so called spectral noise voltages or spectral noise currents. It applies that[2]

$$\tilde{v}_{Rn} = \sqrt{S(f)} \qquad (4.1.62)$$

with unit $\dfrac{V}{\sqrt{Hz}}$ or, respectively,

$$\tilde{i}_{Rn} = \sqrt{S(f)} \qquad (4.1.63)$$

with unit $\dfrac{A}{\sqrt{Hz}}$. Then it applies to the root mean square that:

---

[2] The line above the symbol refers to the root mean square, the subscript R to a random signal and subscript n to the spectral normalization (to 1 Hz). We also talk about normalized spectral voltages and currents.

$$\tilde{v} = \sqrt{\int\limits_{-\infty}^{+\infty} \tilde{v}_{Rn}^2(f)\mathrm{d}f} \qquad (4.1.64)$$

We can then use the WIENER–KHINTCHINE theorem to calculate the spectral power density:

$$S(\omega) = \frac{1}{2\pi} \int\limits_{-\infty}^{+\infty} ACF(\tau)\,\mathrm{e}^{-j\omega\tau}\,\mathrm{d}\tau \qquad (4.1.65)$$

For the back transformation, it results correspondingly that:

$$ACF(\tau) = \int\limits_{-\infty}^{+\infty} S(\omega)\,e^{j\omega\tau}\,\mathrm{d}\omega \qquad (4.1.66)$$

The FOURIER transform of ACF will be represented by spectral power density.

Example 4.4:   Autocorrelation Function of White Noise

Noise, where the power density over the frequency is constant, is called white noise:

$$S(\omega) = P_0 \qquad (4.1.67)$$

where P_0 is a constant. In order to arrive at ACF, the constant power density has to be FOURIER-transformed and applies the FOURIER correspondence:

$$1 \Leftrightarrow \delta(\tau) \qquad (4.1.68)$$

Thus, ACF becomes

$$ACF(\tau) = P_0\,\delta(\tau) \qquad (4.1.69)$$

Figure 4.1.7 represents ACF and the spectral power density of the white noise. It is clearly visible that the white noise is completely uncorrelated.

$$ACF(\tau \neq 0) = 0$$

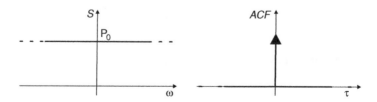

Figure 4.1.7   Spectral power density function and autocorrelation function of white noise

For a frequency range of $-\infty$ to $+\infty$, a constant power density would require infinite energy and thus is not possible. In practice, the white noise of all systems and sensors has a limited bandwidth.

Example 4.5:   Noise Power of Bandwidth-Limited White Noise

As bandwidth limit we assume a bandpass in the frequency range between f_1 and f_2. For the noise power density of the outgoing signal, it thus applies that:

$$S(f) = P_{0n} \quad \text{for } f_1 \leq f \leq f_2 \text{ and } -f_1 \leq f \leq -f_2 \qquad (4.1.70)$$

and otherwise

$$S(f) = 0 \qquad (4.1.71)$$

or applying rectangular function re:

$$S(f) = P_{0n}\left[\text{re}\left(\frac{f-f_0}{B}\right) + \text{re}\left(\frac{f+f_0}{B}\right)\right] \qquad (4.1.72)$$

with bandwidth $B = f_2 - f_1$ and centre frequency $f_0 = \dfrac{f_1 + f_2}{2}$.

The following FOURIER correspondences are required for calculating ACF:

$$\text{re}\left(\frac{f}{B}\right) \Leftrightarrow B\,\text{si}\,(\pi B\tau) \qquad (4.1.73)$$

and

$$g(f-f_0) \Leftrightarrow g(f)e^{-j2\pi f_0 \tau} \qquad (4.1.74)$$

ACF thus becomes

$$ACF(\tau) = P_{0n}\left[B\,\text{si}(\pi\tau B)\,e^{-j2\pi f_0\tau} + B\,\text{si}(\pi\tau B)\,e^{j2\pi f_0\tau}\right] \qquad (4.1.75)$$

The exponential functions can be summarised as

$$ACF(\tau) = 2\,P_{0n}B\,\text{si}(\pi\tau B)\cos\,(2\pi f_0\tau) = P_0\,\text{si}(\pi\tau B)\cos(2\pi f_0\tau) \qquad (4.1.76)$$

with

$$P_0 = 2\,P_{0n}B \qquad (4.1.77)$$

ACF is different outside $\tau = 0$ (Figure 4.1.8). This means that the limitation of the bandwidth causes internal interdependencies in the stochastic noise signal. The noise voltage cannot adopt any order values.

According to Equation 4.1.53, the noise power of the bandwidth-limited noise corresponds exactly to ACF at point $\tau = 0$:

$$P = P_0 = 2P_{0n}B \qquad (4.1.78)$$

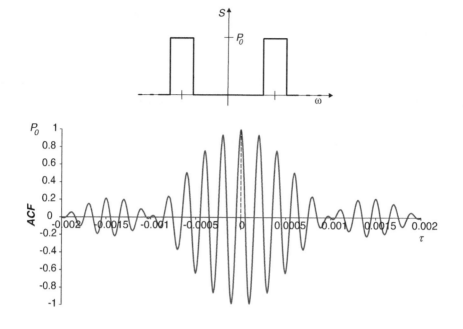

Figure 4.1.8   Autocorrelation function of bandwidth-limited white noise: (a) bandwidth-limited power density; (b) autocorrelation function of (a)

It should be noted that the factor 2 is caused by the distribution of the noise to the positive and negative frequency ranges (two-sided spectrum). Almost all physical descriptions of the noise assume a one-sided spectrum. The above formula has to be corrected correspondingly (the 2 is dropped).

As for white noise, the noise power simply results from multiplying the spectral noise power density with the bandwidth of an ideal rectangular filter. It has been proven advantageous to convert real filters to rectangular filters. We define the so-called equivalent noise bandwidth B_{eq} in a way that the noise power of a real filter is equal to that of an ideal rectangular filter:

$$B_{\mathrm{eq}} = \frac{1}{2\pi|H_{\mathrm{max}}|^2} \int\limits_{-\infty}^{+\infty} |H(j\omega)|^2 \, d\omega \qquad (4.1.79)$$

with transfer function $H(j\omega)$ and maximum H_{max} of the transfer function of a real filter. Thus, it applies for any filter that

$$P = S_0 B_{\mathrm{eq}} \qquad (4.1.80)$$

Example 4.6:   Equivalent Noise Bandwidth of a First-Order Low-Pass

A first-order low-pass has the transfer function

$$H(j\omega) = \frac{1}{1+j\omega\tau} \tag{4.1.81}$$

The equivalent noise bandwidth is then

$$B_{eq} = \frac{1}{2\pi} \int\limits_{-\infty}^{+\infty} \frac{d\omega}{1+\omega^2\tau^2} = \frac{1}{2\tau} \tag{4.1.82}$$

For a one-sided power spectrum we have to integrate Equation 4.1.82 from 0 to ∞ and it becomes:

$$B_{eq} = \frac{1}{4\tau} \tag{4.1.83}$$

Often integrators are used for evaluating the sensor signal. The integrator with integration time t_0 has the following transfer function:

$$H(j\omega) = \frac{\sin t_0\omega}{t_0\omega} \tag{4.1.84}$$

From this, we can calculate the equivalent noise bandwidth, which results in

$$B_{eq} = \frac{1}{2\pi} \int\limits_{-\infty}^{+\infty} \frac{\sin^2 t_0\omega}{t_0^2\omega^2} d\omega = \frac{1}{2t_0} \tag{4.1.85}$$

This relation can also be used if we want to calculate the variance and have to integrate over the spectral power density (averaging).

For calculating the power spectrum of a POISSON-distributed pulse sequence, we can use the CARSON theorem. It states that a pulse sequence is the superimposition of K independent pulses of the same form:

$$x(t) = \sum_{k=1}^{K} a_k f(t-t_k) \tag{4.1.86}$$

with random amplitude a_k of the k-th pulse at random time t_k. Function $f(t)$ describes the form of the pulses. The mean pulse rate is

$$v = \lim_{T\to\infty} \frac{K}{T} \tag{4.1.87}$$

The mean pulse amplitude and the square mean value are

$$\bar{a} = \lim_{T\to\infty} \frac{1}{K} \sum_{k=1}^{K} a_k \tag{4.1.88}$$

and

$$\overline{a^2} = \lim_{T \to \infty} \frac{1}{K} \sum_{k=1}^{K} a_k^2 \tag{4.1.89}$$

The mean value of $x(t)$ is

$$\overline{x(t)} = v\,a \int_{-\infty}^{+\infty} f(t)\,dt \tag{4.1.90}$$

If now random time t_k follows a POISSON distribution, according to the CARSON theorem in Equation 4.1.86, it applies for a one-sided capacity spectrum that:

$$S(\omega) = 2v\,\overline{a^2}|F(\omega)|^2 + 4\pi\,\overline{x(t)}^2\delta(\omega) \tag{4.1.91}$$

The FOURIER-transform of $f(t)$ is $F(\omega)$. Here, we assume that the distribution of amplitudes a_k is unrelated to the distribution of pulse time t_k. In reference [1], we provide a general form of the CARSON theorem for processes that do not follow a POISSON distribution. The right-hand term of the equation corresponds to the offset of the pulse sequence and thus to the expected value (characterised by a Dirac pulse at point $\omega = 0$). However, the expected value is constant for ergodic processes. Therefore, this term can be dropped when analysing the noise.

4.1.6   Noise Analysis of Electronic Circuits

In sensors, there are often several noise sources that superimpose each other. For the sensor output, we can then only measure a single noise voltage or noise current that contains all noise components. For calculating parameters of these noise variables from the individual noise contributions, we use methods of the small-signal analysis of electronic circuits.

If noise sources interfere with the measuring system, there is an output noise signal even if there is no input signal. If we know transfer function $H(j\omega)$ of the measuring system, we can construe the noise-contaminated system by interconnecting a noiseless system and noise-equivalent input signals $\tilde{v}_{nx}$ and $\tilde{i}_{nx}$ (Figure 4.1.9a and b). If the input signal itself is noisy, it can be accordingly summed up using the noise-equivalent input value. As the noise powers add up, it applies for the noise voltages, for instance,

$$\tilde{v}_{nx,\text{ges}} = \sqrt{\tilde{v}_{nx1}^2 + \tilde{v}_{nx2}^2} \tag{4.1.92}$$

If $H(j\omega)$ is the transfer function of a noiseless system, the spectral noise power density S_{yy} at the output can be calculated as:

$$S_{yy}(\omega) = |H(j\omega)|^2 S_{xx}(\omega) \tag{4.1.93}$$

with input noise power density S_{xx}.

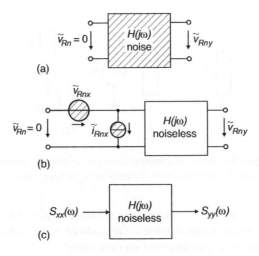

(a)

(b)

(c)

Figure 4.1.9   (a) System with noise sources, (b) noiseless system with noise-equivalent input signals, and (c) transfer of the noise power density of a noiseless system. Noisy components are hatched

---

Example 4.7:   Spectral Noise Power Density of a First-Order Low-Pass

At a low-pass input, we have white noise with a noise power of

$$S_{xx}(\omega) = P_0 \tag{4.1.94}$$

The low-pass has the transfer function

$$H(j\omega) = \frac{1}{1+j\omega\tau} \tag{4.1.95}$$

The squared modulus of the transfer function is

$$|H(j\omega)|^2 = \frac{1}{1+\omega^2\tau^2} \tag{4.1.96}$$

The spectral noise power density at the filter output then is

$$S_{yy}(\omega) = \frac{P_0}{1+\omega^2\tau^2} \tag{4.1.97}$$

For the noise analysis, at first we have to design the small-signal equivalent circuit diagram. There, all noisy components are replaced by their noise-equivalent circuit diagram (see Figure 4.1.9b). The noise-equivalent circuit diagram contains all noise sources as noise voltage or noise current sources. We can use these noise sources as if they were sinusoidal signal sources. Figure 4.1.10 shows examples of noise equivalent circuit diagrams of a resistor. Both are equivalent. Noisy components are presented hatched. If required, the directional arrow can be added to noise current sources or noise voltage sources. It does not have any physical importance, but often simplifies the analysis of the circuit.

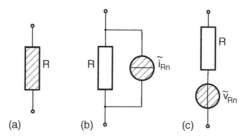

Figure 4.1.10 Noise-equivalent circuit of a resistor: (a) noisy resistor, (b) noiseless resistor with parallel noise current source, (c) noiseless resistor with in-series noise voltage source

Now, we determine a noise output voltage for each noise source. An essential difference to small-signal analysis consists in that all noise sources add up as the sum of squares of the noise power of k noise sources $\tilde{v}_{m,k}$ superimposed on each other:

$$\tilde{v}_{\mathrm{Rn}} = \sqrt{\sum_{k} \tilde{v}_{\mathrm{Rn},k}^2} \qquad (4.1.98)$$

The effective value of the noise voltage at the output is

$$\tilde{v}_{\mathrm{R}} = \sqrt{\int_{f_1}^{f_2} \tilde{v}_{\mathrm{Rn}}^2 \, df} \qquad (4.1.99)$$

Example 4.8: Output Noise of an Infrared Bolometer

Figure 4.1.11 represents an input circuit of a bolometer. At a constant operational voltage V_0, the voltage is decoupled over bolometer resistor R_{B} using field-effect transistor J_1. The field-effect transistor can be represented as current source $g_m v_g$ controlled by gate voltage v_g. Load resistor R_{L} has to have the same temperature as the bolometer resistor. Drain resistor R_{D} completes the transistor stage.

Figure 4.1.12 represents a simplified noise-equivalent circuit diagram. Resistors R_{B} and R_{L} can be summed as they have the same temperature. The noise current is (see Equation 4.2.3 in Section 4.2)

$$i_{\mathrm{Rn,BL}}^2 = \frac{4 k_{\mathrm{B}} T}{R_{\mathrm{B}} \| R_{\mathrm{L}}} \qquad (4.1.100)$$

Leakage current I_{L} of the field-effect transistor is noisy at the gate (current noise, see Section 4.3):

$$i_{\mathrm{Rn},J}^2 = 2\, e\, I_{\mathrm{L}} \qquad (4.1.101)$$

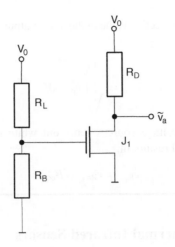

Figure 4.1.11   Input circuit of a bolometer

At the circuit output, there is drain resistor R_D with thermal noise:

$$\tilde{i}^2_{Rn,D} = \frac{4k_B T}{R_D} \qquad (4.1.102)$$

Now we calculate the contributions of all three noise sources to the noise output voltage:

$$\tilde{v}^2_{Rn,BL} = \tilde{i}^2_{Rn,BL} \left(R_B \| R_L\right)^2 g^2_m R^2_D \qquad (4.1.103)$$

$$\tilde{v}^2_{Rn,J} = \tilde{i}^2_{Rn,J} \left(R_B \| R_L\right)^2 g^2_m R^2_D \qquad (4.1.104)$$

$$\tilde{v}^2_{Rn,BL} = \tilde{i}^2_{Rn,D} R^2_D \qquad (4.1.105)$$

which then add up geometrically. The normalised noise voltage at the output thus becomes

$$\tilde{v}_{Rn,a} = R_D \sqrt{\left(\tilde{i}^2_{Rn,BL} + \tilde{i}^2_{Rn,J}\right) \left(R_B \| R_L\right)^2 g^2_m + \tilde{i}^2_{Rn,D}} \qquad (4.1.106)$$

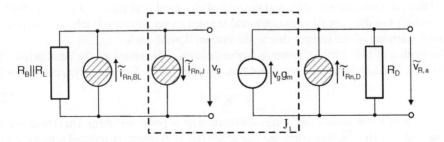

Figure 4.1.12   Noise-equivalent circuit diagram for Figure 4.1.11

Finally, for calculating the effective value of the noise output voltage we have to integrate over the frequency range $f_1 \ldots f_2$:

$$\tilde{v}_{R,a} = \sqrt{\int_{f_1}^{f_2} \tilde{v}_{Rn,a}^2 \, df} \qquad (4.1.107)$$

As all mentioned noise voltage sources represent white noise, solving the integral according to Equation 4.1.80 results in:

$$\tilde{v}_{R,a} = \tilde{v}_{Rn,a} \sqrt{B_{eq}} \qquad (4.1.108)$$

4.2 Noise Source in Thermal Infrared Sensors

4.2.1 Thermal Noise and $\tan \delta$

Similar to the BROWNian motion of the molecules, free charge carriers move arbitrarily in a resistor. The random movements induce voltage changes at the resistor's electrodes. The resulting, measurable, stochastic signal is called thermal noise, also known as JOHNSEN noise or NYQUIST noise or simply as resistance noise. The spectral noise power is unrelated to the resistance:

$$S_{R,th} = 4k_B T \qquad (4.2.1)$$

Here, T is the absolute temperature and k_B the BOLTZMANN constant. This relation is called NYQUIST formula. The resistance noise is normally distributed white noise. The noise voltage can be calculated from the spectral noise power resulting in

$$\tilde{v}_{Rn,th}^2 = S_{R,th} \cdot R = 4 \, k_B T R \qquad (4.2.2)$$

and the noise current resulting in

$$\tilde{i}_{Rn,th}^2 = \frac{S_{R,th}}{R} = \frac{4k_B T}{R} \qquad (4.2.3)$$

Figure 4.2.1 shows the noise current and the noise voltage as a function of the resistance.

The thermal noise of a resistor is an equilibrium process, that is, the above formulas only apply if the resistor is in thermal equilibrium and field-free in the interior. In practice, it has been shown that for a stationary current and the related electrical field there is a quasi-equilibrium in the resistor that does not affect the measurable noise power. We have to take into account the potential warming of the resistor due to the current flow, though.

The motion of the individual charge carriers causes a disruption of the equilibrium. The equilibrium is restored only after relaxation time τ_R. The NYQUIST formula only applies as long as

$$\omega^2 \tau_R^2 \ll 1 \qquad (4.2.4)$$

If this condition is not fulfilled any longer, the thermal noise power will drop. The relaxation time is in the order of 10^{-12} s. Regarding the frequency ranges relevant for thermal sensors, we can always assume white thermal noise.

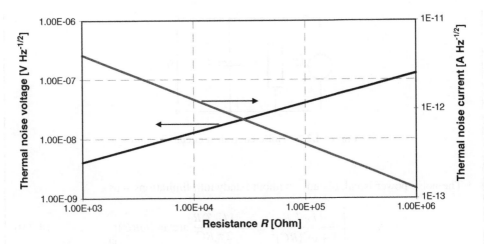

Figure 4.2.1 Thermal resistance of a resistor at $T = 300\,\mathrm{K}$

The presented thermal noise applies to pure ohmic resistors. For complex resistors (impedance Z), the thermal noise is calculated as:

$$\tilde{v}_{Rn,th}^2 = 4\,k_B\,T\,\mathrm{Re}\{\underline{Z}\} \tag{4.2.5}$$

as only the real component of an impedance results in a dissipation and thus in a noise contribution. Purely complex resistors, for example, ideal capacitors, do not produce any noise. However, in practice, there are no ideal capacitors. Due to dielectric losses and leakage currents, real-life capacitors have a loss resistance R_C, that is parallel to capacity C. It can be described using loss angle δ_C:

$$\tan \delta_C = \frac{1}{\omega C R_C} \tag{4.2.6}$$

The loss resistance causes the so-called tan δ noise:

$$\tilde{i}_{Rn,\delta}^2 = 4 k_B T \omega C \tan \delta_C \tag{4.2.7}$$

The tan δ noise is also called dielectric noise.

Example 4.9: Noise of a Lossy Capacitor

Figure 4.2.2 shows a noise-equivalent circuit diagram of a resistor with a capacitor connected in parallel.

The noise output voltage of the circuit in Figure 4.2.2 is

$$\tilde{v}_{Rn,a} = \sqrt{\frac{4 k_B\,T\,R}{1 + \omega^2 \,(RC)^2}} \tag{4.2.8}$$

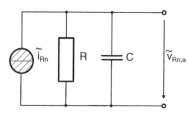

Figure 4.2.2   Noise-equivalent circuit diagram

The noise power is calculated – without bandwidth limitations – as

$$\tilde{v}_{R,a}^2 = \int_0^\infty \frac{4\,k_B\,T\,R}{1+\omega^2(RC)^2}\,df = \frac{4\,k_B\,T\,R}{2\,\pi\,RC}\,\arctan\left(\omega RC\right)\Big|_0^\infty \qquad (4.2.9)$$

The result is the so-called *kTC* noise:

$$\tilde{v}_{R,a}^2 = \frac{k_B\,T}{C} \qquad (4.2.10)$$

or – refering to the name – the variance of the charge on the capacitor:

$$\tilde{Q}_{R,a}^2 = k_B TC \qquad (4.2.11)$$

Without any restrictions of the bandwidth, the effective noise voltage over the capacitor does not depend any longer on the noisy resistor, but exclusively on capacitance and temperature. Thus, the noise current over a capacitor with a capacitance of 1 nF and a temperature of 300 K amounts to approximately 2 µV (see Figure 4.2.3).

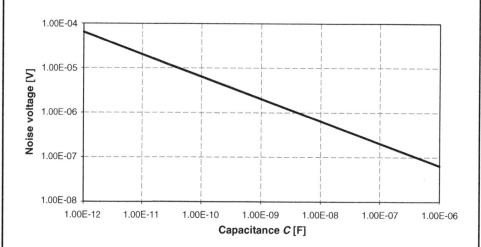

Figure 4.2.3   *kTC* noise of a lossy capacitor at $T = 300\,K$

4.2.2   Current Noise

If charge carriers have to overcome a potential barrier, the energy distribution of the individual charge carriers results in different probabilities of entering another potential and thus in current noise. It is also called shot noise. Such potential barriers are pn-junctions or SCHOTTKY (metal–semiconductor) junctions.

The current is number N of charge carriers q that flow through a cross-sectional area per time unit:

$$I(t) = \sum_{n=1}^{N} q\delta\,(t{-}t_n) \tag{4.2.12}$$

Each individual charge now has to overcome the potential barrier itself and independent of other charge carriers. The number of charge carriers that tunnel through the barrier per time unit follows a POISSON distribution. We can use the CARSON theorem according to Equations 4.1.86 to 4.1.91 to calculate the spectral power density of such a pulse sequence, the pulse amplitude $\bar{a}$ of a single occurrence for a charge can be assumed to be a Dirac pulse and thus

$$\bar{a} = q \tag{4.2.13}$$

with the pulse form

$$F(\delta(t)) = 1 \tag{4.2.14}$$

The product of pulse amplitude $\bar{a}$ and mean pulse rate v then becomes the effective value of current I:

$$\bar{a}\,v = I \tag{4.2.15}$$

The current noise can now be calculated using Equation 4.1.91:

$$S(f) = 2\,q\,I \tag{4.2.16}$$

The same result was already deduced in Section 4.1.3.2 applying a purely mathematical way (Equation 4.1.47). Figure 4.2.4 shows the calculation of an example. Current noise is POISSON-distributed white noise. The above formulas only apply as long as the charges can be described as white noise.

4.2.3   1/f Noise

Particularly in semiconductors, there are fluctuations that occur very frequently for slow processes or low frequencies, respectively, or very seldom for fast processes or high frequencies, respectively. Therefore, this phenomenon is called 1/f-noise. There is no clear explanation regarding the origin of the 1/f-noise. Depending on its occurrence, in conductors or semiconductors, for instance, there are several possible reasons that overlap:

- local temperature fluctuations that cause fluctuations in the thermal equilibrium (especially in thin layers),
- superimposition of several generation–recombination mechanisms in semiconductors,

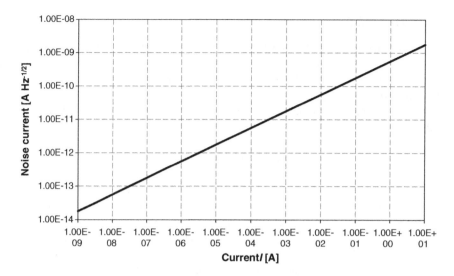

Figure 4.2.4 Current noise of an electron current (charge q = elementary charge e)

- migration of impurities in semiconductors,
- changes of the contact resistance over time (contact noise),
- frequency-related conductivity (e.g. for hopping conduction) and so on.

They all have in common the characteristic curve of the spectral noise power density:

$$S(\omega) \sim \frac{1}{\omega^x} \qquad (4.2.17)$$

with $x = 1 \ldots 2$. $1/f$-noise is also called pink noise.

Example 4.10: 1/f Noise of a Semiconductor Resistor

According to reference [1] it applies to semiconductor resistors that

$$S(f) = \frac{KI^\alpha}{|f|^\beta} \qquad (4.2.18)$$

with parameters $\alpha \approx 2$ and $\beta \approx 1$ and constant K. Equation 4.2.18 applies to the frequency range between 10^{-6} Hz (corresponding to a cycle duration of about 11.5 days!) and 10^6 Hz. In practice, parameters α, β and K are mostly empirically determined. We arrive at the $1/f$-noise power by integrating Equation 4.2.18:

$$\tilde{v}^2_{r,1/f} = \int_{f_1}^{f_2} K\frac{I^2}{f}\,df = K\,I^2\,\ln\frac{f_2}{f_1} \qquad (4.2.19)$$

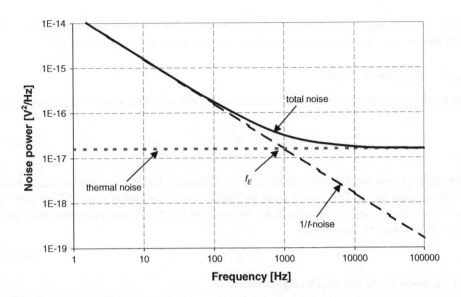

Figure 4.2.5 Noise power density of a semiconductor resistor. $R = 1\,\text{k}\Omega$ ($\tilde{v}_{Rn,th} = 4\,\text{nV}/\sqrt{\text{Hz}}$); edge frequency $f_E = 1\,\text{kHz}$

Both frequencies f_1 and f_2 are unknown at first, but can be approximated. The upper detection limit of the $1/f$-noises corresponds to the intersection of spectral power density and thermal noise of the resistor (Figure 4.2.5). This frequency is called edge frequency f_E. The total noise power is calculated by adding up both noise powers:

$$\tilde{v}_R^2 = \tilde{v}_{R,1/f}^2 + \tilde{v}_{R,th}^2 \tag{4.2.20}$$

Thus we can achieve sufficient precision if the upper frequency limit f_2 is set at a much larger value than the corner frequency f_E, for instance:

$$f_2 \approx 10 \cdot f_E \tag{4.2.21}$$

The lower frequency limit f_1 is determined by observation time t_O. The following rule of thumb applies:

$$f_1 \approx \frac{1}{4\,t_O} \tag{4.2.22}$$

The observation time t_O has to have a value that is large enough for the measured noise power to remain approximately constant.

The 1/f-noise must not be mistaken for flicker noise. Flicker noise is proportional to $1/f^2$. Drift phenomena can also be treated as 1/f-noise, but are usually not stationary.

4.2.4   Radiation Noise

The emission of photons of a blackbody approximately follows a POISSON distribution [2]:

$$\sigma^2 = \bar{n} \left[\frac{e^{\frac{h\nu}{k_B T}}}{e^{\frac{h\nu}{k_B T}} - 1} \right] \tag{4.2.23}$$

with the expected number of photons $\bar{n}$. The right-hand factor in Equation 4.2.23 is the so-called boson factor. This results from the probability of the energy state of BOSE particles (bosons) – photons are also bosons. For a temperature $T < 500$ K and a wavelength range of 0.3 to 30 µm, we can assume that

$$h\nu \ll k_B T \tag{4.2.24}$$

and the boson factor can thus be approximated as

$$\frac{e^{\frac{h\nu}{k_B T}}}{e^{\frac{h\nu}{k_B T}} - 1} \approx 1 \tag{4.2.25}$$

It is possible to arrive at an exact POISSON distribution for this approximation. The expected number of photons each with energy $h\nu$ accordingly corresponds to the PLANCK distribution function:

$$\bar{n} = \frac{1}{e^{\frac{h\nu}{k_B T}} - 1} \tag{4.2.26}$$

We use the CARSON theorem in Equation 4.1.91 to calculate the power spectrum. The time function of the emitted photon flux with optical frequency ν is

$$f_\nu(t) = \sum_{n=1}^{N} h\nu \, \delta(t - t_n) \tag{4.2.27}$$

with the stochastically distributed time t_n. The pulse amplitude is

$$\bar{a} = h\nu \tag{4.2.28}$$

and the pulse form is that of a DIRAC delta function:

$$F(\delta(t)) = 1 \tag{4.2.29}$$

The mean emission rate $\bar{N}$ per wavelength results from PLANCK'S radiation law in Equation 2.3.2 and the emitting area A_{BB}:

$$\bar{N} = \frac{A_{BB} M_\nu}{h\nu} \tag{4.2.30}$$

with the frequency-specific exitance M_ν of a blackbody. Taking into consideration the boson factor in Equation 4.2.23, it has – in turn – to be inserted in Equation 4.2.30 [3]:

$$\bar{N} = \frac{A_{BB} M_\nu}{h\nu} \frac{e^{\frac{h\nu}{k_B T}}}{e^{\frac{h\nu}{k_B T}} - 1} \tag{4.2.31}$$

The frequency-specific power spectrum (related to optical frequency v) then becomes

$$S_{BB,v}(f) = 2A_{BB}(hv)^2 \frac{M_v}{hv} \frac{e^{\frac{hv}{k_BT}}}{e^{\frac{hv}{k_BT}}-1} = 2A_{BB}\,hv\,M_{vS} \frac{e^{\frac{hv}{k_BT}}}{e^{\frac{hv}{k_BT}}-1} \qquad (4.2.32)$$

The result is independent of the frequency (related to the electrical frequency f), that is, the radiation noise is white noise. The integration over the applied optical wavelength range v_1 to v_2 results in the power spectrum of the radiation:

$$S_{BB} = 2A_{BB}\int_{v_1}^{v_2} hv\,M_v \frac{e^{\frac{hv}{k_BT}}}{e^{\frac{hv}{k_BT}}-1}\,dv \qquad (4.2.33)$$

A known solution of the integral for the total radiation ($v_1 = 0$; $v_2 = \infty$) is:

$$S_{BB} = 8A_{BB}\sigma k_B T^5 \qquad (4.2.34)$$

Example 4.11: Signal-to-Noise Ratio of the Total Radiation of a Blackbody

The signal-to-noise ratio (*SNR*) is the ratio between the signal and the effective value of the noise. For the total radiation of a blackbody, the signal can be calculated using the STEFAN–BOLTZMANN law according to Equation 2.3.4. Applying Equation 4.2.34, *SNR* thus becomes

$$SNR = \frac{\sigma T^4 A_{BB}}{\sqrt{8\sigma k_B T^5 A_{BB} B_{eq}}} = \sqrt{\frac{\sigma A_{BB}}{8k_B B_{eq}}}T^3 \qquad (4.2.35)$$

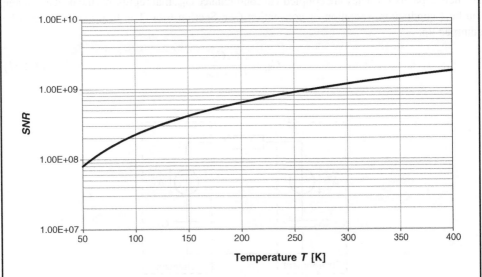

Figure 4.2.6 Signal-to-noise ratio *SNR* of the total radiation of a blackbody; $B = 1\,\text{Hz}$; $A_{BB} = 1\,\text{cm}^2$

Figure 4.2.6 shows the *SNR* for a bandwidth of 1 Hz and a radiating area of 1 cm². Assuming a technically feasible bandwidth of 10 kHz, at 300 K, the resulting *SNR* becomes approximately 140 dB.

4.2.5 Temperature Fluctuation Noise

The temperature of a body describes the average kinetic energy per particle and motion type. As both change over time, there are temperature fluctuations around a constant mean value (temperature T), even in thermal equilibrium. The variance of these temperature fluctuations are according to reference [3]

$$\overline{(\Delta T)^2} = \frac{k_B\, T^2}{C_{th}}$$

(4.2.36)

Here, k_B is BOLTZMANN'S constant and C_{th} the specific heat capacity of the body. $\overline{(\Delta T)}$ can be assumed to be the effective value of the temperature change. Therefore, Equation 4.2.36 can be interpreted as temperature fluctuation noise (or temperature variation noise). Temperature fluctuation noise is white noise, that is, it has a noise power density that is independent of the frequency.

If both bodies are in thermal equilibrium and are then thermally coupled, there will be a noise power flux due to these temperature fluctuations. Figure 4.2.7 shows the thermal model of a sensor element and its surroundings.

The surrounding has temperature T_1 and heat capacity C_1; and the sensor has temperature T_2 and heat capacity C_2. They are coupled via conductance G_{th}, that represents the heat exchange via heat conduction, radiation or convection (see Section 6.1). The heat capacities can be summed up:

$$C_{th} = \frac{C_1 C_2}{C_1 + C_2}$$

(4.2.37)

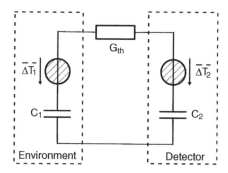

Figure 4.2.7   Thermal model for calculating the temperature fluctuation noise of a sensor element and its surroundings

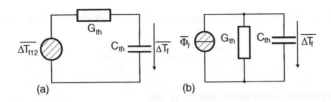

Figure 4.2.8 Thermal equivalent circuits for the arrangement in Figure 4.2.8: (a) equivalent circuit derived by summarising circuit elements, (b) with source substitution from the equivalent circuit transformed in (a)

Both heat capacities are in series. Assuming that heat capacity C_1 of the surroundings is much larger than C_2 of the sensor, it applies that

$$C_{th} \approx C_2 \qquad (4.2.38)$$

According to Equation 4.2.36, it results for the temperature fluctuation that

$$\overline{\Delta T_{12}^2} = \frac{k_B T_1^2}{C_1} + \frac{k_B T_2^2}{C_2} \approx \frac{k_B T_2^2}{C_{th}} \qquad (4.2.39)$$

Figure 4.2.8 shows two thermal equivalent circuits derived from Figure 4.2.7 that can be converted into each other using the methods of source substitution. Index f points out that here we have effective values that are normalised to a bandwidth of 1 Hz, that is, spectral values.

The thermal equivalent circuit in Figure 4.2.8a can be used to directly derive the mean temperature fluctuation:

$$\overline{\Delta T_f^2} = \frac{G^2}{G^2 + \omega^2 C_{th}^2} \overline{\Delta T_{f12}^2} \qquad (4.2.40)$$

As the temperature fluctuation noise is white noise, it applies for the noise power density that

$$\overline{\Delta T_{f12}^2} = \frac{\overline{\Delta T_{12}^2}}{B_{eq}} \qquad (4.2.41)$$

For determining the equivalent noise bandwidth B_{eq}, we can immediately apply Equation 4.1.83, as the system in Figure 4.2.8 is a first-order low-pass:

$$B_{eq} = \frac{1}{4\tau_{th}} \qquad (4.2.42)$$

with τ_{th} being the thermal time constant

$$\tau_{th} = \frac{C_{th}}{G_{th}} \qquad (4.2.43)$$

Using Equations 4.2.39 and 4.2.41, the spectral noise power density then becomes

$$\overline{\Delta T_{f12}^2} = 4\tau_{th}\overline{\Delta T_{12}^2} = \frac{4k_B T_2^2}{G_{th}} \tag{4.2.44}$$

The wanted temperature fluctuation density is now

$$\overline{\Delta T_f^2} = \frac{4k_B G_{th}}{G_{th}^2 + \omega^2 C^2} T_2^2 \tag{4.2.45}$$

If we know the temperature fluctuation density, Figure 4.2.8b can be used to calculate the corresponding noise flux $\overline{\Phi_f}$:

$$\overline{\Phi_f^2} = (G_{th}^2 + \omega^2 C^2)\overline{\Delta T_f^2} = 4k_B G_{th} T_2^2 \tag{4.2.46}$$

The noise flux only depends on the temperature of the sensor and the thermal conductance between surroundings and sensor element. It is independent of the heat capacity and thus of the material qualities of the sensor element. The noise power can be determined by multiplying Equation 4.2.46 by bandwidth B:

$$\overline{\Phi_{R,T}^2} = 4\,k_B G_{th} T^2 B \tag{4.2.47}$$

Thermal conductance G_{th} describes the heat conduction ($G_{th,G}$), for example, due to the suspension of a sensor element, and the radiaton of the sensor area ($G_{th,R}$). Both thermal conductances have to be summed up as they are 'driven' by the same temperature difference, that is, they are parallel:

$$G_{th} = G_{th,G} + G_{th,R} \tag{4.2.48}$$

For a greybody, the thermal conductance by heat radiation can be calculated using the STEFAN–BOLTZMANN law according to Equation 2.3.4:

$$G_{th,R} = \frac{d\Phi}{dT} = 4\varepsilon\sigma T^3 A_S \tag{4.2.49}$$

Figure 4.2.9 constitutes an example of the thermal conductance by radiation. Inserting the thermal conductance by heat radiation $G_{th,R}$ (Equation 4.2.14) for the thermal element G_{th} in Equation 4.2.47, we get the noise flux by radiation:

$$\overline{\Phi_{R,T}^2} = 16\varepsilon k_B \sigma T^5 A_S B \tag{4.2.50}$$

Comparing this result to the effects of the radiation noise in Equation 4.2.34, we can see that they are identical, with the exception of a factor 2. Equation 4.2.50 can also be expressed as the noise power being composed of the noise of the absorbed radiation (surroundings with

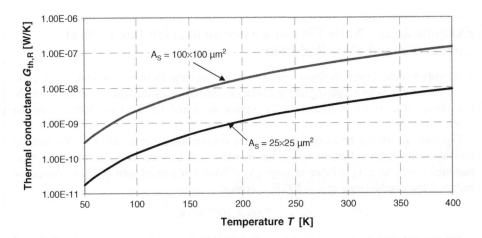

Figure 4.2.9   Conductance $G_{th,R}$ due to radiation between a sensor element with area A_S and its surroundings ($\varepsilon = 1$)

temperature T_1) and the noise of the emitted radiation (sensors with temperature T_2) (emission and absorption are statistically uncorrelated):

$$\overline{\Phi^2_{R,T}} = [S_{BB}(T_1) + S_{BB}(T_2)]B \tag{4.2.51}$$

Therefore, it is possible to approximately express the noise flux due to heat radiation as

$$\overline{\Phi^2_{R,T}} = 8\,\varepsilon\,k_B\,\sigma\,(T_1^5 + T_2^5)\,A_S\,B \tag{4.2.52}$$

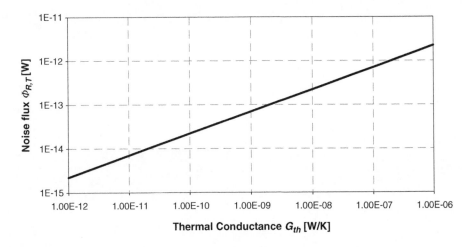

Figure 4.2.10   Noise flux $\Phi_{R,T}$ in a microbolometer bridge due to heat conduction and heat radiation; $T = 300\,\mathrm{K}$; $B = 1\,\mathrm{Hz}$

Example 4.12:   Noise Flux in a Microbolometer Due to Heat Conduction and Heat Radiation

A microbolometer bridge is always situated in the vacuum. Between the sensor areas and the environment (sensor housing and read-out circuit), there are undesired, but unavoidable heat fluxes due to the radiation of the sensor area and the heat conduction via the contact feet.

Equation 4.2.49 can be used to calculate the conductance due to radiation. We have to take into consideration that the sensor has an upper and a lower side. This means that the assumed sensor area A_S is twice as large ($A_S = 2A_P$). For a pixel area of $25 \times 25\,\mu m$, for instance, the conductance at $T = 300$ K becomes:

$$G_{th,S} = 7.66 \times 10^{-9}\ \text{W/K}$$

For modern sensors, conductance $G_{th,L}$ of the contact feet is between 1×10^{-8} W/K and $1 \times 10{-}7$ W/K. Figure 4.2.10 shows noise flux $\Phi_{R,T}$ in relation to the conductance.

References

1. Blum, A. (1996) *Elektronisches Rauschen (Electronic Noise)*, Teubner Verlag, Stuttgart.
2. Dereniak, E.L. and Boreman, G.D. (1996) *Infrared Detectors and Systems*, John Wiley & Sons Ltd, Chichester.
3. Kruse, P.W. and Skatrud, D.D. (1997) *Uncooled Infrared Imaging Arrays and Systems; Semiconductors and Semimetals*, vol. **47**, Academic Press, New York.

5

Sensor Parameters

The characteristics of a sensor can be described by its parameters. These parameters can be divided into two major groups. One of them describes thermal resolution, such as responsivity, and the other spatial resolution, such as the modulation transfer function.

5.1 Responsivity

5.1.1 Introduction

The responsivity describes the change of an output variable's value due to a change in the input variable that causes this change [1]. For thermal IR sensors, this input value is radiant flux Φ_S and the output variable is output voltage V_S or output current I_S. The definition of voltage responsivity R_V thus becomes

$$R_V = \frac{\Delta V_S}{\Delta \Phi_S} \tag{5.1.1}$$

with the change of output voltage ΔV_S being caused by the change in radiant flux $\Delta \Phi_S$. The unit of the voltage responsivity is V/W. There is an analogous definition of current responsivity R_I for sensors with current I_S as the output variable:

$$R_I = \frac{\Delta I_S}{\Delta \Phi_S} \tag{5.1.2}$$

with the unit A/W.

According to the kind of radiation, we distinguish between blackbody responsivity – also known as black responsivity (Section 5.1.2) – and spectral responsivity (Section 5.1.3). For sensor applications in temperature measuring devices, the temperature-related responsivity of a blackbody is very important. This responsivity R_T is the slope of the signal transfer function (*SiTF*) (Section 5.1.4). For sensors with several sensitive areas (picture elements or pixels), in addition, it is important to state the uniformity (Section 5.1.5).

Thermal Infrared Sensors: Theory, Optimisation and Practice   Helmut Budzier and Gerald Gerlach
© 2011 John Wiley & Sons, Ltd

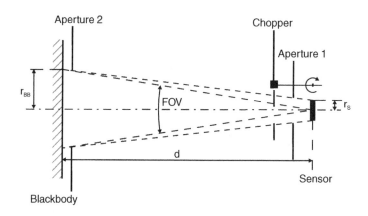

Figure 5.1.1   Measuring arrangement for determining black responsivity

5.1.2   Black Responsivity

If we use a blackbody as radiation source for determining the responsivity of a sensor we talk about black responsivity. This does not take into account the absorption characteristics of the sensor. For calculating the radiant flux, we always start from total radiation. Figure 5.1.1 shows the measurement arrangement. The sensor area with radius r_S and the blackbody area with radius r_{BB} are located at distance d. We use aperture stops (aperture 1 and 2) to determine the solid angle of the sensor. We distinguish between constant light and alternating light measurements. The latter requires an additional chopper for the temporal modulation of the radiation amplitude. The simplest case is a rotating multi-bladed wheel at constant frequency f_{CH}.

For steady-state measurements, we work with a radiant flux that is constant in time. For two different blackbody temperatures, the output DC voltage of the sensor is measured and the responsivity is calculated according to Equation 5.1.1.

Example 5.1:   Measuring the Responsivity of a Microbolometer Array

We will measure the responsivity of a microbolometer array for a 30 °C blackbody temperature T_{BB}. For this purpose, we arrange the microbolometer array at a distance of $d = 50$ mm in front of a blackbody. It is common practice to determine the parameters of arrays for an f-number of $F = 1$. Thus, the radius of the blackbody becomes

$$r_{BB} = 25 \text{ mm} \tag{5.1.3}$$

We assume a pixel size of $25 \times 25 \text{ μm}^2$ (pixel area $A_S = 6.25 \times 10^{-10} \text{ m}^2$). As the pixel area is much smaller than the emitting area ($A_{BB} = 1.96 \times 10^{-3} \text{ m}^2$) we can assume a differentially small area element. The radiant flux at the pixel then becomes

$$\Phi_S = \frac{\varepsilon \sigma T_{BB}^4}{4 k^2 + 1} A_S \tag{5.1.4}$$

For determining the responsivity we use the radiation difference between two different blackbody temperatures. Therefore, we do not need to calculate the radiant flux exchanged between blackbody and sensor. Due to the calculation of the difference, the sensor's self-radiation is not included. We use the ambient temperature (e.g. 25 °C) as the second blackbody temperature. For $\varepsilon = 0.98$, the radiation difference we are looking for is

$$\Delta \Phi_S = \Phi_S(30°C) - \Phi_S(25°C) = 58.7 \, \text{nW} - 54.9 \, \text{nW} = 3.8 \, \text{nW}. \tag{5.1.5}$$

The following values are measured as sensor output voltages:

$$V_S(25°C) = 2.4567 \, \text{V} \tag{5.1.6}$$

and

$$V_S(30\,°C) = 2.4601 \, \text{V} \tag{5.1.7}$$

The measured values also include the bias voltage of the pixel. The voltage responsivity thus is

$$R_V(30°C) = \frac{3.4 \, \text{mV}}{3.8 \, \text{nW}} = 894\,737 \, \text{V/W}. \tag{5.1.8}$$

The alternating light responsivity is defined for sinusoidal variables. Thus, Equation 5.1.1 becomes:

$$R_V = \frac{\tilde{V}_S}{\tilde{\Phi}_S} \tag{5.1.9}$$

with the effective values of change of radiant flux $\tilde{\Phi}_S$ and the effective value of change of output voltage $\tilde{V}_S$. If we use a chopper for the temporal modulation of the radiation, the result often is not a sinusoidal radiation. For a rotating multi-bladed wheel chopper, we get an approximately rectangular modulation of the radiation. In this case, we use the fundamental harmonic as the sinusoidal signal. The effective value of the radiation's fundamental harmonic is calculated from the radiation difference between chopper and blackbody. The output voltage can then be determined with a frequency-selective voltmeter, for instance. Here, we also measure the fundamental harmonic.

Example 5.2:   Measuring the Responsivity of a Pyroelectric Sensor

We will measure the black responsivity of a pyroelectric sensor with a pixel area of $A_S = 1 \, \text{mm}^2$ and a field of vision *FOV* of 28°. Typically, the following measuring conditions apply:

- Chopper frequency $f_{Ch} = 10 \, \text{Hz}$.
- Blackbody temperature $T_{BB} = 500 \, \text{K}$.
- Sensor temperature $T_S = 25\,°C$ (ambient temperature).

A second blackbody temperature is not given as the sensor sees the chopper during the chopper-closed phase and this temperature is then the second blackbody temperature. The chopper has ambient temperature. The distance between sensor and emitter is $d = 50$ mm. The radius of the emitting blackbody surface is thus

$$r_{BB} = 12.5 \text{ mm} \qquad (5.1.10)$$

Even in this case, we can assume a differentially small area element and it applies that

$$\Phi_S = \varepsilon \sigma T^4 A_S \sin^2\left(\frac{FOV}{2}\right) \qquad (5.1.11)$$

In the chopper-open phase, the sensor sees the blackbody ($\varepsilon = 0.98$):

$$\Phi_S(500 \text{ K}) = 385.9 \,\mu\text{W} \qquad (5.1.12)$$

In the chopper-closed phase, the sensor sees the chopper blade that has ambient temperature ($\varepsilon = 0.98$):

$$\Phi_S(300 \text{ K}) = 50.0 \,\mu\text{W} \qquad (5.1.13)$$

The modulated radiant flux is approximately rectangular. A rectangle signal can be presented as a FOURIER series:

$$\Phi(t) = \Phi_{PP}\left\{\frac{1}{2} + \frac{2}{\pi}\left[\cos(\omega_0 t) + \frac{1}{3}\cos(3\omega_0 t) + \ldots\right]\right\} \qquad (5.1.14)$$

with

$$\omega_0 = 2\pi f_{Ch} \qquad (5.1.15)$$

and the peak-to-peak flux

$$\Phi_{PP} = \Phi_S(500 \text{ K}) - \Phi_S(300 \text{ K}) = 335.9 \,\mu\text{W} \qquad (5.1.16)$$

The amplitude of fundamental harmonic ω_0 is

$$\hat{\Phi}_S = \frac{2}{\pi}\Phi_{PP} = 213.8 \,\mu\text{W} \qquad (5.1.17)$$

or the effective value we are looking for:

$$\tilde{\Phi}_S = \frac{\hat{\Phi}_S}{\sqrt{2}} = 151.2 \,\mu\text{W} \qquad (5.1.18)$$

The choppered radiant flux has a steady component:

$$\Phi_0 = \frac{1}{2}\Phi_{PP} = 167.9 \,\mu\text{W} \qquad (5.1.19)$$

For a pyroelectric sensor, this steady radiation component does not generate a signal, but increases the average sensor temperature. For a constant light-sensitive sensor, it only generates an output signal, but this will be filtered out during the frequency-selective measurement. It should be noted that it does not correspond to the mean temperature between blackbody and chopper surface. In this example, the mean radiant flux corresponds to a blackbody temperature of about 332 K.

The effective value of the output voltage is measured with a selective voltmeter at a frequency of 10 Hz:

$$\tilde{V}_S = 15.3 \, \text{mV} \tag{5.1.20}$$

Now we can calculate the voltage responsivity:

$$R_V = \frac{15.3 \, \text{mV}}{151.2 \, \mu\text{W}} = 101.2 \, \text{V/W} \tag{5.1.21}$$

When stating the responsivity, we often present the measuring conditions as $R_V(T_{BB}, f_{Ch})$. This means that the above measuring result becomes:

$$R_V(500 \, \text{K}, 10 \, \text{Hz}) = 101.2 \, \text{V/W} \tag{5.1.22}$$

Thermal sensors show a typical frequency dependency of responsivity (see Chapter 6). This is represented in Figure 5.1.2 for constant light- and alternating light-sensitive sensors. For the latter, it applies that:

$$R_V(f) = \frac{R_0}{\sqrt{1 + (\omega \tau_{th})^2}} \tag{5.1.23}$$

with voltage responsivity R_0 for $f = 0$ and thermal time constant τ_{th}. For alternating light-sensitive sensors (particularly pyroelectric sensors), it applies that:

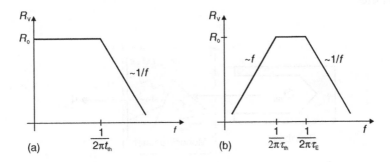

Figure 5.1.2 Typical frequency curve of the responsivity: (a) constant light-sensitive sensors, and (b) pyroelectric sensors

$$R_V(f) = \frac{\omega R_0}{\sqrt{1+(\omega\,\tau_E)^2}\sqrt{1+(\omega\,\tau_{th})^2}} \qquad (5.1.24)$$

with electric time contant τ_E.

5.1.3   Spectral Responsivity

Spectral responsivity R_λ describes the wavelength dependency of a sensor's responsivity:

$$R_\lambda = \frac{\Delta V_S}{\Delta \Phi_\lambda} \qquad (5.1.25)$$

with monochromatic radiation Φ_λ with wavelength λ.

The measuring principle of spectral responsivity corresponds to that of black responsivity; only the radiation source is different. Radiation sources we can use are, amongst others,

- monochromators,
- tunable lasers,
- optical filters.

Figure 5.1.3 shows how monochromatic radiation is generated using a blackbody. In the monochromator, the incident black radiation Φ_{BB} is diffracted with a grating or a prism. We then use an exit slit to select the desired wavelength and direct it to the sensor. Radiation flux Φ_λ is measured with a calibrated sensor. Often we only state a curve that is normalised to the maximum spectral responsivity.

If we are only interested in the spectral responsivity of a specific wavelength or a small wavelength range, we can also use tunable lasers or optical filters.

5.1.4   Signal Transfer Function

The *SiTF* describes the relationship between the temperature of a blackbody and the output voltage of the sensor. In addition to the sensor parameters, the *SiTF* also includes optogeometrical parameters.

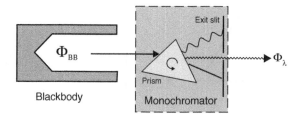

Figure 5.1.3   Generation of monochromatic radiation using a prism-monochromator and a blackbody as radiation source

Radiant flux Φ_S corresponding to one pixel is the product of the radiation power and the area of the pixel A_P:

$$\Phi_S(T) = \frac{A_P \, \tau_{opt}}{4 \, F^2 + 1} \, M_{BB}(T) \qquad (5.1.26)$$

with the transmission τ_{opt} of the optical path. Exitance M_{BB} is wavelength-dependent:

$$M_{BB}(T) = \int_{\lambda_1}^{\lambda_2} M_{\lambda BB}(T) \, d\lambda \qquad (5.1.27)$$

According to PLANCK'S radiation law, the formula for spectral exitance is

$$M_{\lambda BB}(T) = \frac{c_1}{\lambda^5} \frac{1}{\exp\left(\dfrac{c_2}{\lambda \, T}\right) - 1} \qquad (5.1.28)$$

with radiation constants c_1 and c_2, wavelength λ and wavelength ranges λ_1 and λ_2. For $\lambda_1 = 0$ and $\lambda_2 \to \infty$, the established solution of this equation is the STEFAN–BOLTZMANN law:

$$M_{BB}(T) = \sigma \, T^4 \qquad (5.1.29)$$

Including the wavelength-dependent absorption of the pixel, the above equations result in radiant flux Φ_A absorbed by the pixel:

$$\Phi_A(T) = \frac{A_P \, \tau_{opt}}{4 \, F^2 + 1} \int_{\lambda_1}^{\lambda_2} \alpha(\lambda) \, M_{\lambda BB}(T) \, d\lambda \qquad (5.1.30)$$

Example 5.3:   Radiant Flux between Blackbody and Sensor Pixel

Only the radiant flux exchanged between blackbody and pixel results in a signal. For calculating the output voltage, we have therefore to determine the sensor's own temperature (operating point T_0). The measured signal is the difference signal between blackbody temperature and the sensor's own temperature. Figure 5.1.4 shows the flux difference in relation to a sensor temperature of 30 °C:

$$\Delta\Phi_A = \Phi_A(T) - \Phi_A(30°C) \qquad (5.1.31)$$

The absolute radiant flux $\Phi_A(30\,°C)$ of the blackbody, impinging on the pixel, amounts to 43.7 nW ($\lambda = 8 \ldots 14\,\mu m$) or 117.3 nW ($\lambda = 0 \ldots \infty$).

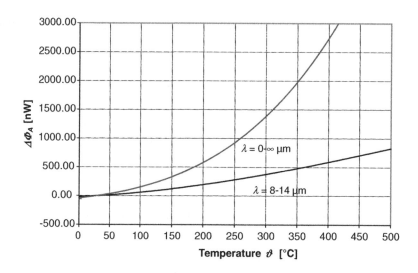

Figure 5.1.4 Radiant flux difference in relation to a sensor temperature of 30 °C. Pixel area $A_P = 35 \times 35\,\mu m^2$; f-number $F = 1$; absorbance $\alpha = 1$; transmittance $\tau = 1$

If we multiply difference flux by voltage responsivity we get voltage difference ΔV, which is the output signal of the pixel:

$$\Delta V = R_V \cdot \Delta \Phi_A \qquad (5.1.32)$$

The temperature-dependent voltage difference ΔV is identical to $SiTF$:

$$SiTF(T) = \frac{R_V\,A_P\,\tau_{opt}}{4\,F^2 + 1} \int_{\lambda_1}^{\lambda_2} \alpha(\lambda)\,M_{\lambda BB}(T)\,d\lambda \qquad (5.1.33)$$

The responsivity regarding temperature thus is the increase of $SiTF$:

$$R_T = \frac{\partial SiTF(T)}{\partial T} \qquad (5.1.34)$$

Responsivity R_T depends on the blackbody temperature. This dependence is included in the exitance:

$$R_T = \frac{\partial SiTF(T)}{\partial T} = \frac{R_V\,A_P\,\tau_{opt}}{4\,F^2 + 1} \int_{\lambda_1}^{\lambda_2} \alpha(\lambda)\,\frac{\partial M_{\lambda BB}(T)}{\partial T}\,d\lambda \qquad (5.1.35)$$

For the differential exitance, it applies that

$$\frac{\partial M_{\lambda BB}}{\partial T} = M_{\lambda BB}(T)\,\frac{c_2}{\lambda\,T^2}\,\frac{1}{1 - \exp\left(-\dfrac{c_2}{\lambda\,T}\right)} \qquad (5.1.36)$$

It is often sufficient to approximate Equation 5.1.36:

$$\frac{\partial M_{\lambda BB}}{\partial T} \approx M_{\lambda BB}(T) \frac{c_2}{\lambda T^2} \tag{5.1.37}$$

For values $\lambda T < 3100\ \mathrm{K\,\mu m}$, the deviation caused by this approximation is smaller than 1 %.

Example 5.4:   Differential Exitance

For calculating the differential exitance, we integrate Equation 5.1.36 over wavelength range from λ_1 to λ_2:

$$I_M = \frac{\partial M_{BB}}{\partial T} = \int_{\lambda_1}^{\lambda_2} \frac{\partial M_{\lambda BB}}{\partial T}\, d\lambda \tag{5.1.38}$$

The differential exitance states how much the radiant flux of an object changes with changing temperatures. The larger this value is, the higher is the responsivity that can be reached. Figure 5.1.5 shows the differential exitance for different wavelength ranges. Further values are provided in Example 5.8. The differential exitance of the total radiation ($\lambda_1 = 0$ and $\lambda_2 = \infty$) can be easily calculated using the STEFAN–BOLTZMANN law:

$$\frac{\partial M_{BB}}{\partial T} = 4\,\sigma\,T^3 \tag{5.1.39}$$

This is the maximum differential exitance.

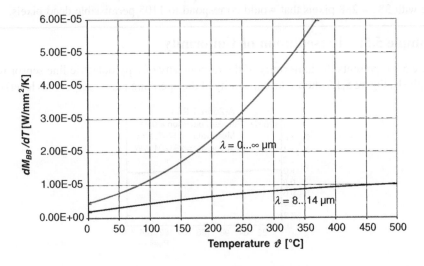

Figure 5.1.5   Differential exitance

5.1.5   Uniformity

Due to tolerances that apply to the manufacturing of sensor arrays, the individual sensor elements (pixels) of a sensor have different specific values and parameters. The uniformity describes the difference between the individual pixels. In particular, this refers to the operating point (offset values, bias, DC output level) and the responsivity.

The uniformity is also called fixed-pattern noise. This term is misleading, though, as the uniformity is not a noise, but is determined.

We have to state the uniformity separately for the responsivity and for the operating point as both are uncorrelated. The uniformity includes the statement of:

- mean value,
- standard deviation,
- minimum and maximum values,
- limit values,
- inoperative pixels.

For arrays with only a small number of pixels, the individual pixel values can be stated in a table. For large pixel numbers, the uniformity is represented graphically:

- as a line graph for row-like arrangements,
- as a histogram.

Inoperative pixels and pixels with values outside the stated tolerances are called 'dead pixels'. Dead pixels are presented in tables. Clusters of dead pixels are mainly classified according to size and position (middle or margin of the array). Uniformity and dead pixels are the most important quality characteristics of an array and to a large extent determine the price. The upper allowable limit for the number of dead pixels is usually 1%. For a microbolometer bridge with 384×288 pixels that would correspond to 1105 permissible dead pixels.

Example 5.5:   Presentation of Uniformity

Figure 5.1.6 presents the uniformity of the responsivity of a pyroelectric line sensor with 256 pixels. Figure 5.1.7 shows the uniformity of the operating point (output bias level) of a

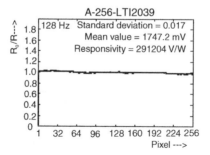

Figure 5.1.6   Uniformity of the responsivity of a pyroelectric line sensor (By courtesy of DIAS Infrared GmbH, Dresden, Germany)

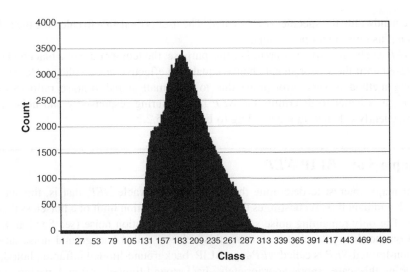

Figure 5.1.7 Uniformity of the output bias level of a microbolometer array with 640×480 pixels; output voltage range linearly divided into 512 classes

microbolometer array – that is, the offset values – in a histogram. The spread of the offset values limits the dynamic range of the array and is often larger than the signal from objects at ambient temperature.

5.2 Noise-Equivalent Power *NEP*

Noise-equivalent power (*NEP*) is the radiant flux that generates a signal-to-noise ratio of Unity at the sensor output:

$$NEP = \frac{\tilde{v}_{R}}{\Delta V_{S}} \Delta \Phi_{S} \qquad (5.2.1)$$

or

$$NEP = \frac{\tilde{v}_{R}}{R_{V}} \qquad (5.2.2)$$

with sensor noise voltage $\tilde{v}_{R}$, voltage responsivity R_{V} and output voltage change ΔV_{S}, that is caused by the change in radiant flux $\Delta \Phi_{S}$. *NEP* has the unit W. If we relate *NEP* to the normalised noise voltage $\tilde{v}_{R,n}$, we talk about the specific *NEP* (*NEP**, also *NEP* star):

$$NEP^{*} = \frac{\tilde{v}_{R,n}}{R_{V}} \qquad (5.2.3)$$

The unit of the specific *NEP* is $\mathrm{W}/\sqrt{\mathrm{Hz}}$. Stating the *NEP* is useful for sensors with integrated electronics that determines the noise bandwidth. For sensors where the user electronics defines the noise bandwidth, the specific *NEP* is stated. The user is then able to calculate the achieved *NEP* using his own signal processing.

The *NEP* can be directly measured. For this purpose, the temperature of a blackbody has to be increased until the sensor signal change equals the effective noise voltage. However, this measuring method is very error-prone due to the small signal-to-noise ratio $(SNR \approx 1)$. Therefore, it is better to determine the *NEP* by measuring responsivity and noise voltage and subsequently calculating it according to Equation 5.2.2.

Example 5.6: BLIP-*NEP*

It is of major interest to determine the minimum achievable *NEP*, that is, the smallest detectable radiant flux. It constitutes the theoretical resolution limit of a noiseless thermal sensor. The only remaining noise source is then the radiation noise (see Section 4.2.5). Based on the theory of photon sensors, this noise is called background noise and the radiation-limited *NEP* is called NEP_{BLIP} (BLIP: background-limited infrared photodetection or in this case, more appropriately, background-limited infrared performance). Inserting the noise flux according to Equation 4.2.52, we arrive at

$$NEP_{\text{BLIP}} = \sqrt{8\varepsilon k_B \sigma \left(T_S^5 + T_0^5\right) A_S B} \qquad (5.2.4)$$

with sensor temperature T_S and background temperature T_0 (ambient temperature). BLIP-*NEP* thus only depends on the temperature, the sensor area and the noise bandwidth. If we assume the sensor area and noise bandwidth to be determined by the sensor design, BLIP-*NEP* can only be reduced by lowering the temperature.

Cooling the sensor to temperature $T_S = T_0$ only results in an improvement by factor $\sqrt{2}$. This means that also the sensor environment, that is, the background, has to be cooled. It is important that the sensor's *FOV* is preserved. Equation 5.2.4 becomes

$$NEP_{\text{BLIP}}(FOV) = \sqrt{8\varepsilon k_B \sigma T_{BB}^5 A_S B \sin^2 \frac{FOV}{2}} \qquad (5.2.5)$$

with

- cooling the sensor (reduction by factor $\sqrt{2}$),
- using a cooled aperture stop,
- reducing the noise flux of the sensor's *FOV*,
- where T_{BB} is the temperature of the signal source (blackbody as object).

Using cooled sensors with cooled aperture stop can decrease BLIP-*NEP*:

$$NEP_{\text{BLIP}}(FOV) = \frac{NEP_{\text{BLIP}}(FOV = \pi)}{\sin^2 \dfrac{FOV}{2}} \qquad (5.2.6)$$

Another way of reducing BLIP-*NEP* is to restrict the wavelength range of the radiation (see Equation 4.5.11).

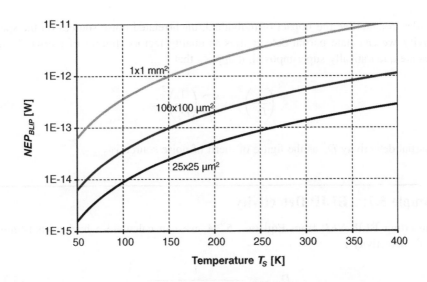

Figure 5.2.1 BLIP-*NEP* of a thermal sensor. Parameter: sensor area A_S; $T_S = T_0$; $\varepsilon = 1$; $B = 1$ Hz

Figure 5.2.1 shows the BLIP-*NEP* of a thermal sensor in relation to the temperature T_S and with the sensor area A_S as parameter.

5.3 Detectivity

Detectivity *D* is a variable derived from the noise-equivalent power. It is the reciprocal value of *NEP* and characterises the *SNR*:

$$D = \frac{1}{NEP} \tag{5.3.1}$$

If we include detectivity as a sensor parameter we are able to qualitatively evaluate different sensors. Larger detectivity values correspond to qualitatively better sensors. Usually, we use specific detectivity D^* (also, *D* star):

$$D^* = \frac{\sqrt{A_S}}{NEP^*} = \frac{\sqrt{A_S}}{\tilde{v}_{R,n}} R_V \tag{5.3.2}$$

The unit of the specific detectivity is m Hz$^{1/2}$/W. Sometimes particularly in the United States the non-SI unit Jones is used (1 Jones = 1 cm Hz$^{1/2}$/W).

The specific detectivity is calculated from the measured quantities, responsivity R_V and noise voltage $\tilde{v}_{R,n}$.

In order to underline the impact of individual, uncorrelated noise sources on the specific detectivity, we can state partial detectivities as quality factors (figures of merit). As noise sources are quadratically superimposed, it applies that

$$\left(\frac{1}{D^*}\right)^2 = \sum_i \left(\frac{1}{D_i^*}\right)^2 \qquad (5.3.3)$$

with partial detectivity D_i^* as the figure of merit of noise source $\tilde{v}_{\mathrm{Rn,i}}$.

Example 5.7: BLIP Detectivity

By inserting BLIP-*NEP* from Equation 5.2.4, we can calculate a background-limited specific detectivity[1]:

$$D_{\mathrm{BLIP}}^* = \frac{1}{\sqrt{8\varepsilon k_{\mathrm{B}}\sigma(T_{\mathrm{S}}^5 + T_0^5)}} \qquad (5.3.4)$$

Figure 5.3.1 shows the background-limited specific detectivity as a function of the sensor temperature.

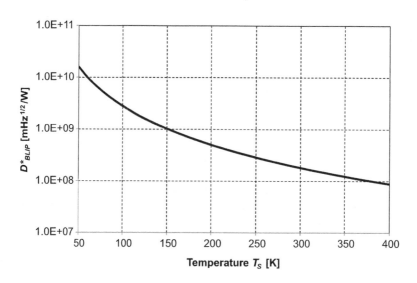

Figure 5.3.1   BLIP detectivity of a thermal sensor: $T_{\mathrm{S}} = T_0$; $\varepsilon = 1$; $B = 1\,\mathrm{Hz}$

---

[1] In order to achieve the specific BLIP-*NEP*, Equation (5.2.4) still has to be divided by $\sqrt{B}$.

5.4   Noise-Equivalent Temperature Difference

The temperature resolution of a sensor is described by the noise-equivalent temperature difference (*NETD*). It is the temperature difference ΔT in the object that generates a *SNR* of 1:

$$NETD = \frac{\tilde{v}_R}{\Delta V_S} \Delta T \qquad (5.4.1)$$

At the same time this definition equation states the measurement instructions. Figure 5.4.1 shows a common measuring scenario. Temperature difference ΔT is the difference between the background with temperature T_0 and the object with temperature T_{BB}:

$$\Delta T = T_{BB} - T_0 \qquad (5.4.2)$$

Background temperature T_{BB} is at the same time the reference temperature for the *NETD* specification. Signal difference ΔV_S results from the temporal mean values of the noisy voltage:

$$\Delta V_S = \bar{V}_{BB} - \bar{V}_0 \qquad (5.4.3)$$

Noise voltage $\tilde{v}_R$ is the effective value of the sensor noise voltage. For a *SNR* of 1, the object is hardly visible in the measuring scenario as the peak-to-peak signal of the noise is substantially larger than the differential signal[2]. This means that *NETD* is the absolutely smallest detectible temperature difference in this scenario. For measuring *NETD* in this scenario we select a substantially larger temperature difference. The resulting measurement deviation will be calculated later on in Example 5.8.

For measuring *NETD*, it is advantageous to use the differential form of Equation 5.4.1:

$$NETD = \frac{\tilde{v}_R}{\dfrac{\partial V_S}{\partial T}} \qquad (5.4.4)$$

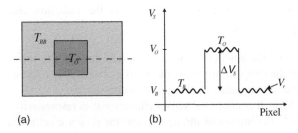

(a)   (b)

Figure 5.4.1   (a) Scenario for measuring *NETD* and (b) voltage curve along the dashed cutting line

---

[2] For a GAUSSIAN distribution of the noise voltage, the peak-to-peak signal is approximately six times larger than the effective value.

In the nomonator of Equation 5.4.4 we now find the responsivity regarding temperature at background temperature T_{BB} (Equation 5.1.34):

$$NETD = \frac{4\,F^2+1}{A_{\mathrm{P}}\,\tau_{\mathrm{opt}}} \cdot \frac{1}{\varepsilon \displaystyle\int_{\lambda_1}^{\lambda_2} \alpha(\lambda)\frac{\partial M_{\lambda\mathrm{BB}}}{\partial T}\,\mathrm{d}\lambda} \cdot \frac{\tilde{v}_{\mathrm{R}}}{R_{\mathrm{V}}} \qquad (5.4.5)$$

or using the specific detectivity:

$$NETD = \frac{4\,F^2+1}{\sqrt{A_{\mathrm{P}}}\,\tau_{\mathrm{opt}}} \cdot \frac{1}{\varepsilon \displaystyle\int_{\lambda_1}^{\lambda_2} \alpha(\lambda)\frac{\partial M_{\lambda\mathrm{BB}}}{\partial T}\,\mathrm{d}\lambda} \cdot \frac{\sqrt{B}}{D^*} \qquad (5.4.6)$$

or using *NEP*:

$$NETD = \frac{4\,F^2+1}{A_{\mathrm{P}}\,\tau_{\mathrm{opt}}} \cdot \frac{1}{\varepsilon \displaystyle\int_{\lambda_1}^{\lambda_2} \alpha(\lambda)\frac{\partial M_{\lambda\mathrm{BB}}}{\partial T}\,\mathrm{d}\lambda} \cdot NEP \qquad (5.4.7)$$

It is quite easy to interpret the three terms in Equation 5.4.7:

- Left-hand term $\frac{4F^2+1}{A_{\mathrm{P}}\,\tau_{\mathrm{opt}}}$: this term contains only optogeometrical parameters. It is indirectly proportional to projected solid angle ω_1. In Equations 5.4.5 to 5.4.7, this term is determined by f-number F of the applied optics and by pixel size A_{P}. *NETD* is nearly proportional to the square of f-number F.
- Middle term $\dfrac{1}{\varepsilon \int_{\lambda_1}^{\lambda_2} \alpha(\lambda)\frac{\partial M_{\lambda\mathrm{BB}}}{\partial T}\,\mathrm{d}\lambda}$: This term describes the radiation characteristics of the measuring object and the absorption of the radiation through the sensor. That means that *NETD* depends on the temperature of the measuring object! This reference temperature (the so-called background temperature) always has to be stated.
- Right-hand term *NEP*: It contains the sensor characteristics responsivity and noise. We may also have to include the impact of the device and the sensor electronics.

In addition to the sensor characteristics, the integral over the differential exitance has a large influence on *NETD*. The solution of the integral depends on the temperature of the measuring object and the selected wavelength range. We always assume the measuring object to be a greybody with the emission coefficient being independent of the wavelength. In order to take into account the effect of the propagation path, for example, the optics, transmission τ_{opt} is included in the integration. The integral over the differential exitance thus becomes

$$I_M = \int_{\lambda_1}^{\lambda_2} \alpha(\lambda)\, \tau_{opt}(\lambda)\, \frac{\partial M_{\lambda BB}(T)}{\partial T}\, d\lambda \tag{5.4.8}$$

Using Equation 5.4.8, it applies for $NETD$ that

$$NETD = \frac{4\,F^2 + 1}{A_P}\, \frac{NEP}{\varepsilon\, I_M} \tag{5.4.9}$$

The simplest solution of integrals I_M results for the total radiation ($\lambda_1 = 0$; $\lambda_2 = \infty$) and $\alpha = \tau = 1$:

$$I_{M\infty} = 4\sigma T^3 \tag{5.4.10}$$

Equation 5.4.10 is the maximum of the differential exitance and thus determines the minimum of $NETD$.

Example 5.8:   BLIP-*NETD*

Using value $I_{M\infty}$ and with $\varepsilon = \alpha = 1$, we achieve the best, that is, minimum, $NETD$ for BLIP-*NEP* in Equation 5.2.4:

$$NETD_{BLIP} = (4F^2 + 1)\sqrt{\frac{k_B B}{\sigma\, T_S\, A_S}} \tag{5.4.11}$$

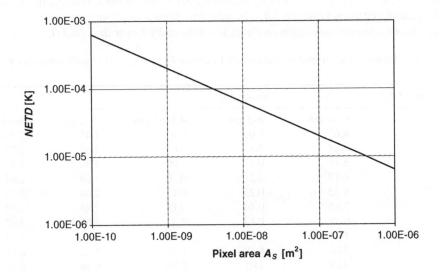

Figure 5.4.2   BLIP-*NETD*: f-number $F = 1$; bandwidth $B = 1$ Hz; sensor temperature = background temperature $T_S = 300$ K

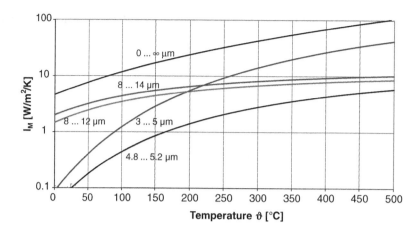

Figure 5.4.3   Integral over differential exitance I_M ($\alpha = \tau_{opt} = 1$). Parameter: wavelength ranges

Figure 5.4.2 shows BLIP-*NETD* as a function of the pixel size. It should be noted that pixel area A_P is included twice in sensor area A_S (front and back!).

Figure 5.4.3 and Table 5.4.1 present the solution of integral I_M in Equation 5.4.8 over blackbody temperature T_{BB} for some important spectral ranges. Usually, *NETD* is stated for a background temperature T_0 of 27 °C (300 K) and an f-number of 1. If the sensor does not work under these conditions, the starting value of the measuring range and the smallest possible f-number is given as background temperature.

NETD is measured either – as already described – according to Equation 5.4.1 and the measuring scenario in Figure 5.4.1 or by measuring *SiTF* and the noise voltage. In the latter, *NETD* is calculated using Equation 5.4.4. In order to achieve a small measuring deviation, we usually choose a temperature difference that is substantially larger than *NETD*.

Table 5.4.1   Values of the integral over differential exitance I_M ($\alpha = \tau_{opt} = 1$) for different wavelength ranges

Temperature (°C)	I_M	[W/(m² K)]			
ϑ_{BB}	0–∞ μm	3–5 μm	4.8–5.2 μm	8–12 μm	8–14 μm
0	4.62	0.09	0.05	1.46	2.00
10	4.88	0.12	0.07	1.65	2.23
20	5.71	0.17	0.09	1.84	2.47
27	6.13	0.21	0.10	1.98	2.64
30	6.32	0.23	0.13	2.04	2.71
50	7.65	0.40	0.18	2.45	3.19
80	9.99	0.82	0.32	3.06	3.92
100	11.8	1.23	0.44	3.47	4.39
200	24.0	5.50	1.41	5.32	6.51
300	42.7	14.1	2.80	6.76	8.13
400	69.2	26.7	4.35	7.84	9.33
500	105	41.9	5.86	8.65	10.2

Example 5.9:   Measuring *NETD*

For determining *NETD* for a background temperature T_0 of 27 °C we can use an object temperature T_{BB} of 30 °C, for instance. In the following we want to analyse the measuring deviation that is caused by the non-linear *SiTF* if temperature difference ΔT – when measuring *NETD* – is substantially larger than *NETD*. Relating Equations 5.4.1 and 5.4.4 to each other, it results relative measuring deviation ΔF:

$$\Delta F = \frac{\dfrac{\tilde{v}_R}{\Delta U_S}}{\dfrac{\Delta T}{\dfrac{\tilde{v}_R}{\dfrac{\partial V_S}{\partial T}}}} = \frac{\Delta T}{\Delta V_S}\frac{\partial V_S}{\partial T} \tag{5.4.12}$$

For the output voltage at total radiation, it applies that

$$V_S = K\sigma T^4 \tag{5.4.13}$$

and thus:

$$\frac{\partial V_S}{\partial T} = 4K\sigma T^3 \tag{5.4.14}$$

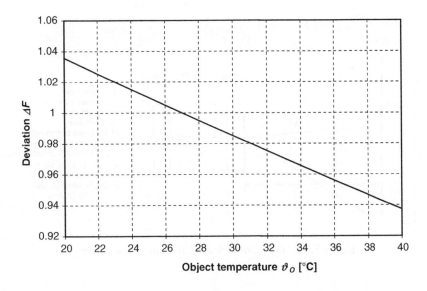

Figure 5.4.4   Relative deviation ΔF of *NETD* when measuring with difference quotients

with K being a system-specific constant. Equation 5.4.12 now becomes

$$\Delta F = 4 \frac{T_{BB} - T_0}{T_{BB}^4 - T_0^4} T_0^3 \qquad (5.4.15)$$

Figure 5.4.4 shows measuring deviation ΔF. For object temperatures above the background temperature, we will measure too small, that is, too optimistic *NETD* values.

5.5 Optical Parameters

The measuring field of a radiation sensor is determined by aperture stops and an optical imaging system (optics, lens). Aperture stops define the field of view of the sensor, but do not produce an optical image. Therefore aperture stops are usually only used for measuring in order to determine sensor characteristics (see Figure 5.1.1). The optics projects a measuring spot to the sensor. Figure 5.5.1 uses a line sensor to show the geometric relations. The optics are described by the principal planes H and H' and by focal points F and F' (or focal lengths f and f'). In addition, we need f-number F or entrance pupil diameter D_0 to calculate the optics. In order to simplify the relations, in the following we will assume projection from the infinite $(R \rightarrow \infty)^3$. Then the image plane is situated in focal plane F'.

In thermography, the measuring field is called the scene. The object to be measured is located in the scene. Anything that does not belong to the object is considered to be part of the background. The size of scene L is determined by the field of view *FOV* and distance R of the scene to the optics, more precisely to principal plane H of the optics:

$$L = 2 R \tan \frac{FOV}{2} \qquad (5.5.1)$$

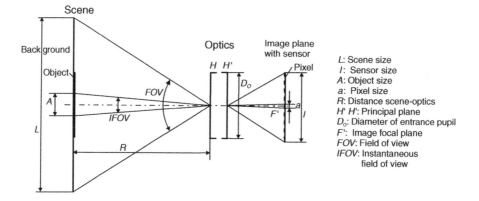

Figure 5.5.1 Optogeometrical relations in a sensor array

---

[3] In practice, it is sufficient if R is substantially larger than f'.

In the scene, size A of a pixel is given by instantaneous field of view (*IFOV*) of the pixel:

$$A = 2R\tan\frac{IFOV}{2} \qquad (5.5.2)$$

FOV results from sensor size l and focal length f' of the optics:

$$FOV = 2\arctan\frac{l}{2f'} \qquad (5.5.3)$$

or for *IFOV* of the pixel with width a:

$$IFOV = 2\arctan\frac{a}{2f'} \qquad (5.5.4)$$

If the sensor is not square, all optical parameters have to be given for both horizontal and vertical direction. For marking, we use the indices h and v, such as FOV_h and FOV_v. The field of view is often given in radians. The following conversions apply:

$$1\,\text{rad} = \frac{180°}{\pi} \approx 57.296°$$

or

$$1° = \frac{\pi}{180}\,\text{rad} \approx 17.45\,\text{mrad}$$

The irradiance of a sensor area behind an optics depends on the amount of light that passes the optics. The luminosity of the optics is stated with f-number F:

$$F = \frac{f'}{D_0} \qquad (5.5.5)$$

Figure 5.5.2 shows a schematic presentation of the path of the radiation from an object-side area element dA_1 through entrance pupil EP of the optics to the image-side area element dA_2. The radiant flux on the entrance pupil is, according to Equation 3.2.28

$$\Phi_1 = A_1\pi L_1\Omega_0\sin^2\varphi_1 \qquad (5.5.6)$$

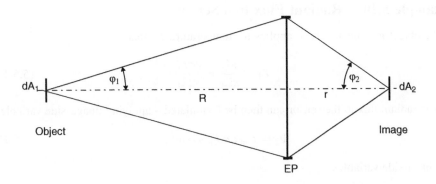

Figure 5.5.2   Path of the radiation from object to image

The expression $\sin\varphi_1$ is called the aperture. In general, it applies to the so-called numerical aperture NA that

$$NA = n_1\sin\varphi_1 \qquad (5.5.7)$$

with n_1 being the refractive index of the transmission medium ($n_1 = 1$ for air). For the image-side radiation flux that impinges on sensor element dA_2, it applies that:

$$\Phi_2 = A_2\pi L_2\Omega_0\sin^2\varphi_2 \qquad (5.5.8)$$

If we neglect the transmission losses in the optics, object-side radiation flux Φ_1 has to be identical to image-side radiation flux Φ_2:

$$A_1 L_1\sin^2\varphi_1 = A_2 L_2\sin^2\varphi_2 \qquad (5.5.9)$$

For the optical imaging, ABBE's sine condition applies:

$$\frac{A_1}{A_2} = \frac{\sin^2\varphi_2}{\sin^2\varphi_1} \qquad (5.5.10)$$

This means that also the radiance of image and object has to be identical:

$$L_1 = L_2 \qquad (5.5.11)$$

Taking into account the optical transmission losses τ_{opt} it thus applies for the irradiance that

$$E_2 = \frac{\Phi_2}{A_2} = \pi\,\tau_{opt}L_1\Omega_0\sin^2\varphi_2 \qquad (5.5.12)$$

For a given f-number F, Equation 5.5.12 becomes

$$E_2 = \frac{\pi\,\tau_{opt}\,L_1\,\Omega_0}{4\,F^2 + 1} \qquad (5.5.13)$$

Example 5.10:   Radiant Flux to a Sensor

If the object is a blackbody, it applies for temperature T_1, that

$$L_1 = \frac{M_S}{\pi\Omega_0} = \frac{\varepsilon\sigma T_1^4}{\pi\Omega_0} \qquad (5.5.14)$$

The radiant flux to the sensor can then be formulated – applying image-side variables:

$$\Phi_2 = \tau_{opt}\,\varepsilon\sigma\,T_1^4\,A_2\sin^2\varphi_2 \qquad (5.5.15)$$

or object-side variables:

$$\Phi_2 = \tau_{opt}\,\varepsilon\sigma\,T_1^4\,A_1\sin^2\varphi_1 \qquad (5.5.16)$$

When calculating the reduced solid angle in Section 3.2, we already assumed that the optics can be represented by its entrance pupil. This is confirmed by the above calculation. The effect that an optics has on the irradiance of a sensor element will be described in the following Example 5.11.

Example 5.11:   Gain Effect of an Optics

An object with diameter d_1 is situated at distance R. For small angles, we use the simplifications $\sin x \approx x$ and $\tan x \approx x$. Then angle φ_0 can be calculated as follows:

$$\sin\varphi_0 \approx \frac{d_1}{2R} \tag{5.5.17}$$

The irradiance without optics can thus be approximated as

$$E_{\text{WithoutLens}} = \pi L_1 \Omega_0 \sin^2 \varphi_0 \tag{5.5.18}$$

The irradiance with optics is:

$$E_{\text{WithLens}} = \pi L_1 \Omega_0 \sin^2 \varphi_1 \tag{5.5.19}$$

If we now determine the ratio of the irradiances, it applies that

$$\frac{E_{\text{WithLens}}}{E_{\text{WithoutLens}}} = \frac{\sin^2 \varphi_1}{\sin^2 \varphi_0} \tag{5.5.20}$$

For small angles φ_1 it applies that

$$\sin\varphi_1 \approx \frac{D_0}{2R} \tag{5.5.21}$$

Equation 5.1.20 now becomes

$$\frac{E_{\text{WithLens}}}{E_{\text{WithoutLens}}} = \frac{D_0^2}{d_1^2} = \frac{A_{\text{EP}}}{A_1} \tag{5.5.22}$$

Using an optics the irradiance of a sensor element increases by the ratio of entrance pupil area A_{EP} to object area A_1. Instead of the object, the sensor perceives the enlarged object that is projected to the entrance pupil.

The measuring spot projected by optics onto the sensor is limited in its smallness due to the FRAUNHOFER diffraction. The AIRY disc is considered the smallest projectable light spot. It represents the rotation-symmetrical diffraction image of a circular aperture. The irradiance of the AIRY disc is [2]

$$E(r) = E(0) \frac{2J_1(\omega r)}{\omega r} \qquad (5.5.23)$$

with

$$E(0) = \frac{\Phi_0}{2R^2} \qquad (5.5.24)$$

and

$$\omega = \frac{\pi D_O}{\lambda R} \qquad (5.5.25)$$

with distance R of the sensor plane to the diaphragm, diameter D_O of the aperture, wavelength λ of the radiation, radiant flux Φ_0 through the aperture and first-order BESSEL function J_1. Figure 5.5.3 shows the diffraction pattern. The size of the AIRY disc is assumed to be the first zero of the BESSEL function. It applies for a projection from the infinite ($R=f$) that

$$r_{\text{Airy}} \approx 1.22 \frac{f'\lambda}{D_O} = 1.22\lambda F \qquad (5.5.26)$$

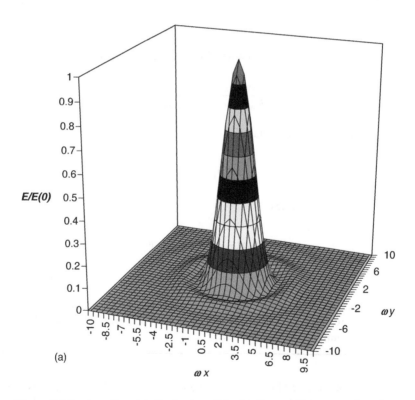

(a)

Figure 5.5.3 AIRY disc: (a) distribution of the irradiance, (b) cross section of (a)

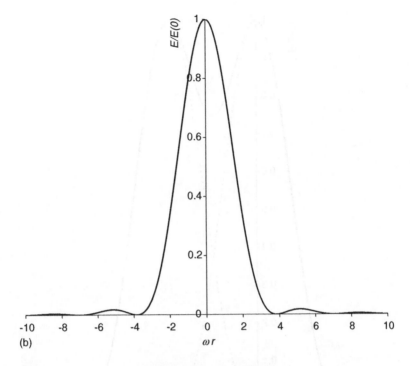

Figure 5.5.3 (*Continued*)

The projection of two incoherent point emitters from the infinite results in two independent AIRY discs that superimpose each other. The distinguishability of the point emitters in the image is considered to be the geometrical resolution limit of an optical system. There are different definitions for the distinguishability. The most famous one is the RAYLEIGH criterion. The RAYLEIGH criterion states that two light points can still be distinguished if the maximum of one light source falls within the minimum of the other light source. The minimum distance then exactly corresponds to the radius of the AIRY disc (Equation 5.5.26). Figure 5.5.4 shows two AIRY discs that barely can be distinguished according to the RAYLEIGH criterion. The minimum between the two is $\omega r \approx 1.9$ and amounts to approximately 0.75.

In order to be able to still resolve the diagram in Figure 5.5.4 using a sensor array, we need at least one pixel in the maximum and one in the minimum. The required pixel distance r_P can thus be calculated as

$$r_P \approx 0.61 \lambda F \tag{5.5.27}$$

A short pixel distance does not increase the spatial resolution of a sensor.

5.6 Modulation Transfer Function

The modulation transfer function (*MTF*) describes the imaging quality of optoelectronic systems, such as infrared sensors. As opposed to the spatial resolution that states the smallest

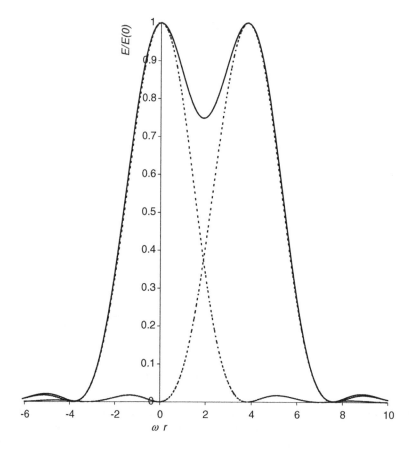

Figure 5.5.4   Superposition of two AIRY discs that can be barely distinguished according to the RAYLEIGH criterion

representable structure, *MTF* describes the transfer of spatial sinusoidal structures in an image as a function of spatial frequency. *MTF* is an objective quantitative measure of the imaging quality. Preconditions for stating the *MTF* are linear systems and incoherent optical imaging.

5.6.1   Definition

The transfer function of an optical system is the optical transfer function (*OTF*) [3–6]. It is defined in analogy to the transfer function $H(j\omega)$ of an electrical system. This means that both functions are compatible with each other and can be converted into each other. Therefore, it is very convenient to use *OTF* for describing a system that includes both optical and electrical components. *OTF* is defined as

$$OTF(f_x,f_y) = \frac{V(f_x,f_y)}{I(f_x,f_y)} \tag{5.6.1}$$

with sinusoidal spatial frequencies f_x or f_y in x- or y-direction, input signal $I(f_x, f_y)$ and output signal $V(f_x, f_y)$. As opposed to the electrical transfer function, *OTF* is normalised to its maximum.

For reasons of clarity in the following we will always use a one-dimensional transfer function (in direction x for a spatial frequency f_x). All statements apply accordingly to the second orthogonal spatial direction y or f_y.

Space-dependent input signal $I(x)$ is assumed to be a spatially sinusoidal-modulated signal (Figure 5.6.1):

$$I(x) = I_0 \cos (2\pi f_x x + \varphi_0) \qquad (5.6.2)$$

Spatial frequency f_x has the unit mm^{-1}. It is common practice to use as the unit of spatial frequency the number of line pairs per millimetre (lp/mm). A line pair is an adjacent pair of light and dark, that is, a spatial period. This means that both units are identical. It applies for spatial frequency f_x that

$$f_x = \frac{1}{T_x} \qquad (5.6.3)$$

with period T_x in mm.

For an identical spatial frequency f_x, the output signal becomes

$$V(x) = V_0 \cos (2\pi f_x x + \varphi_{V0}) \qquad (5.6.4)$$

OTF is a complex function and can therefore be segmented into amplitude (*MTF*) and phase (*PTF*, phase modulation function):

$$OTF\ (f_x) = MTF\ (f_x)\ e^{jPTF\ (f_x)} \qquad (5.6.5)$$

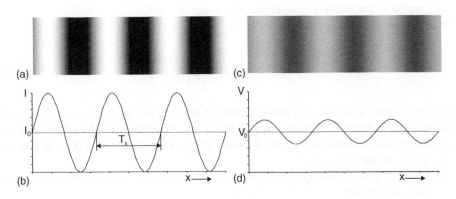

Figure 5.6.1 Input and output signal for $MTF(f_x) = 0.3$. (a) Input signal $I(f_x)$ coded in 256 grey levels, 100% modulation. (b) Intensity of (a) as a one-dimensional diagram. (c) Output signal $V(f_x)$ coded as in a), 30% modulation. (d) Intensity of (c) as a one-dimensional diagram

For infrared radiation sensors we always assume the radiation of a blackbody to be the optical input signal. It always emits incoherent radiation. This means that *OTF* of the optical transfer path becomes real, that is, *PTF* is zero. In the following, we will therefore only consider the amplitude of *OTF*, that is, *MTF*.

If the input signal of a system is a DIRAC delta function, the normalised FOURIER transform of the output signal corresponds to *MTF*. An optical DIRAC delta function is a point source. The image of a point source is called point spread function (*PSF*). This means that *MTF* is the normalised FOURIER transform of the *PSF*:

$$MTF(f_x, f_y) = \left| \frac{F\{PSF(x,y)\}}{F\{PSF(x,y)\}_{f_x=0, f_y=0}} \right| \tag{5.6.6}$$

For the one-dimensional case, the input signal is a line source that we can picture as an array of point sources. The image of a line source is the line spread function (*LSF*). Now *MTF* is the normalised FOURIER transform of the *LSF*:

$$MTF(f_x) = \left| \frac{F\{LSF(x)\}}{F\{LSF(x)\}_{f_x=0}} \right| \tag{5.6.7}$$

Figure 5.6.1 shows a one-dimensional sine function. In this case, we can use the following equation to calculate *MTF*:

$$MTF(f_x) = \frac{V_{max}(f_x) - V_{min}(f_x)}{I_{max}(f_x) - I_{min}(f_x)} \tag{5.6.8}$$

The value of *MTF* at point f_x is also called modulation. *MTF* thus is a representation of the modulation in relation to the spatial frequency. If the modulation equals One, there is an optimum transfer of the spatial structure (input signal = output signal). If the modulation is zero, no spatial information is transmitted. The image is void.

For any input signal, it applies for *MTF* that

$$MTF(f_x) = \frac{1}{MTF_{max}} \left| \frac{F\{V(x)\}}{F\{I(x)\}} \right| \tag{5.6.9}$$

The left-hand term presents the normalisation of *MTF*. In the optics, we normalise to the maximum of *MTF*. When looking at the spatial frequency, all sensors considered here have low-pass characteristics. In the following, we therefore always normalise to *MTF*($f_x = 0$), that is, to the uniformly illuminated image.

For a linear system consisting of subsystems, *MTF* can be calculated as the product of all *MTF*s of all subsystems:

$$MTF = \prod_i MTF_i. \tag{5.6.10}$$

Example 5.12: *MTF* of a Diffraction-Limited Optics

The point spread function of a diffraction-limited optics is the AIRY disc (see Section 5.5). It can be used to derive the one-dimensional *MTF* of an optics for the projection from the infinite [7]:

$$MTF_{\text{Optic}}(f_x) = \frac{2}{\pi}\left[\arccos\left(\lambda\,F\,f_x\right) - \lambda\,F\,f_x\sqrt{1-(\lambda\,F\,f_x)^2}\right] \quad \text{for} \quad \lambda\,F\,f_x \leq 1 \quad (5.6.11)$$

$$MTF_{\text{Optic}}(f_x) = 0 \quad \text{for} \quad \lambda\,F\,f_x > 1 \tag{5.6.12}$$

Figure 5.6.2 shows *MTF* of a diffraction-limited optics. *MTF* becomes zero at point $f_x = 1/(\lambda F)$. It then applies for the period:

$$T_{x0} = \lambda\,F \tag{5.6.13}$$

This value is smaller than the resolution limit defined by Equation 5.5.26. For a distance $r = T_{x0}$, two AIRY discs have almost merged und cannot be distinguished[4]. This means that *MTF* becomes zero. For a spatial frequency with a period that corresponds to AIRY distance r_{Airy} in Equation 5.1.26:

$$f_{\text{Airy}} = \frac{1}{1.22\,\lambda\,F} \tag{5.6.14}$$

MTF amounts to approximately 10%.

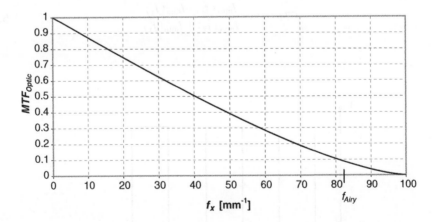

Figure 5.6.2 *MTF* with a diffraction-limited optics: wavelength $\lambda = 10\,\mu\text{m}$; f-number $F = 1$; $f_{\text{Airy}} \approx 82\,\text{mm}^{-1}$

---

[4] According to the SPARROW criterion, two AIRY discs merge at $r = 0.95\,\lambda\,F$.

Until now, we have looked at *MTF* mainly from an optical perspective. From a sensor's view the definition equation (5.6.9) constitutes a normalised responsivity as it applies for the responsivity that

$$R_V = \frac{V(f_x = 0)}{I(f_x = 0)} \tag{5.6.15}$$

Using Equation 5.6.9, it results for *MTF*

$$MTF(f_x) = \frac{R_V(f_x)}{R_V(f_x = 0)} \tag{5.6.16}$$

A spatial frequency of zero means that the sensor is uniformly illuminated. This means that according to Section 5.1, black responsivity R_V is the normalising variable in the nominator of Equation 5.6.16. Thus, *MTF* of a sensor corresponds to the normalised spatial frequency-dependent responsivity.

5.6.2   *Contrast*

Instead of the modulation, sometimes the contrast or the contrast transfer function (*CTF*), is stated. The given contrast always refers to a rectangle grating (Figure 5.6.3). *CTF* is defined as

$$CTF(f_x) = \frac{I_{max}(f_x) - I_{min}(f_x)}{I_{max}(f_x) + I_{min}(f_x)} \tag{5.6.17}$$

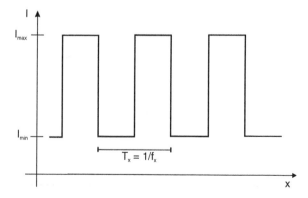

Figure 5.6.3   Contrast due to spatial intensity differences I_{max} and I_{min} for spatial frequency f_x of the fundamental harmonic

with minimum or maximum intensity I of the projected rectangle structure. The slurring of the edge can be neglected when stating the contrast. The contrast has the advantage that it is easy to generate a rectangle structure, which simplifies measurement. It is not possible though to state the contrast of subsystems as the contrasts of systems do not multiplicatively superimpose each other. The contrast can be easily calculated using *MTF*:

$$CTF(f_x) = \frac{4}{\pi}\left[MTF(f_x) - \frac{1}{3}MTF(3f_x) + \frac{1}{5}MTF(5f_x) - \ldots\right] \qquad (5.6.18)$$

5.6.3 *Modulation Transfer Function of a Sensor*

The modulation transfer function of a sensor (sensor *MTF*) is a normalised spatial frequency-dependent responsivity. Sensor *MTF* can be computed by calculating *MTF* of all transfer-relevant subsystems of a sensor. Equation 5.6.10 can be used to calculate sensor *MTF*. According to the operating principle of the sensor, we have to take into account the different subsystems. In the following, we will present several sub-*MTF*s.

5.6.3.1 Geometrical *MTF*

Each pixel has a certain spatial expansion and integrates over its sensor area (Figure 5.6.4). However, we will only look at the one-dimensional case. If the input signal is a DIRAC delta function

$$I(x) = \delta(x) \qquad (5.6.19)$$

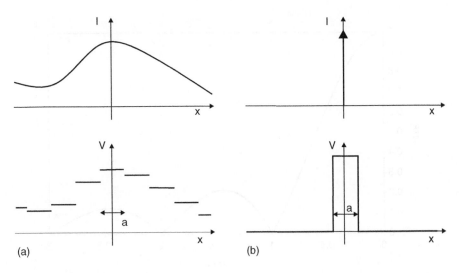

(a)　　　　　　　　　　　　　(b)

Figure 5.6.4　Sensor output signal by sampling with pixel of width a: (a) arbitrary input signal and (b) DIRAC delta function as input signal

the output signal is a *PSF* and represents a rectangle with width a of the pixel (Figure 5.6.4b):

$$V(x) = \text{re}\left(\frac{x}{a}\right) \tag{5.6.20}$$

For calculating *MTF*, we now have to FOURIER-transform $V(x)$. It applies the following correspondence:

$$\text{re}\left(\frac{x}{a}\right) \leftrightarrow a \, \text{si}(a \, \pi f_x) \tag{5.6.21}$$

After normalisation, we arrive at the so-called geometrical *MTF*:

$$MTF_g(f_x) = |\text{si}(a \, \pi f_x)| \tag{5.6.22}$$

Geometrical *MTF* describes the modulation decrease due to the integration of the signal over the pixel area (Figure 5.6.5). The first zero of geometrical MTF_g occurs when the period of spatial frequency f_x is identical to the pixel width a:

$$a = \frac{1}{f_x} \tag{5.6.23}$$

This is the case if we integrate exactly over one period of the sinusoidal input signal.

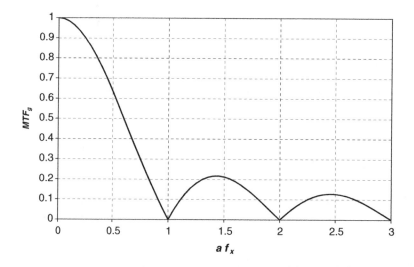

Figure 5.6.5   Geometrical *MTF*

The pixel of a sensor is used to sample the input signal, that is, it is discretised. For an error-free discretisation we have to follow SHANNON'S sampling theorem. It states that the highest frequency contained in the frequency has to be smaller than half the sampling frequency. Applied to spatial sampling, the sampling theorem becomes

$$f_x < \frac{1}{2r} \qquad (5.6.24)$$

with pixel pitch r. The limiting frequency

$$f_{x, \text{Nyquist}} = \frac{1}{2\,r} \qquad (5.6.25)$$

is also called NYQUIST frequency. Now pixel width a cannot be larger than pixel pitch r:

$$a < r \qquad (5.6.26)$$

With this condition, geometrical *MTF* is only physically feasible up to

$$a f_x \le 0.5 \qquad (5.6.27)$$

and thus acquires a minimum value of approximately 64%. This means that geometrical *MTF* is a low-pass. If the signal includes higher frequencies, these are projected in a falsified way. This is called aliasing.

Example 5.13:   Transfer of a Rectangle Signal

A 10 mm long line sensor has 50 μm wide pixels that are arranged in a 50 μm grid. A rectangle signal is projected to it. In order to fulfil the sampling theorem, we use a FOURIER series to synthesise the rectangle signal (Figure 5.6.6a).

The fundamental frequency f_x of the signal is $0.2\,\text{mm}^{-1}$. The FOURIER series is interrupted at the eleventh harmonic ($f_{x,11} = 4.6\,\text{mm}^{-1}$). The result shows a clear overshoot at the edges. Figure 5.6.6b represents optical and geometrical *MTF* of the image of the rectangle signal after low-pass filtering. For optical *MTF*, f-number $F = 3$ and wavelength $\lambda = 10\,\mu\text{m}$ were selected as parameters (Figure 5.6.7). Figure 5.6.6c shows the output signal after sampling by the pixels. We can clearly see the decrease of both edge steepness and overshoot – which is caused by the low-pass behaviour. The asymmetry in Figure 5.6.6c is caused by the asymmetric position of the pixel in relation to the signal. The first pixel area starts at point $x = 0$ (where the sensor starts). The pixel's centre is thus located at $x = 25\,\mu\text{m}$. However, the symmetry point of the signal is situated in $x = 0\,\text{mm}$.

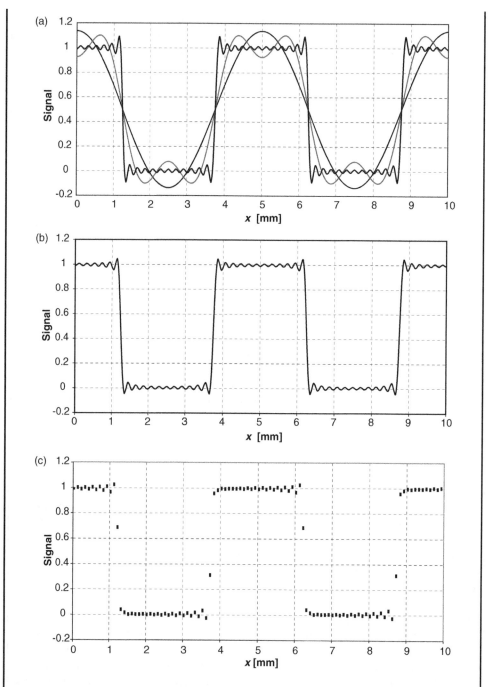

Figure 5.6.6 Effect of *MTF* and sampling of a rectangle signal: (a) Synthesised input signal with $f_x = 0.2 \, \text{mm}^{-1}$. For comparison, the fundamental harmonic as well as the sum of fundamental harmonic and first harmonic are presented, (b) Input signal after impact of optical and geometrical MTF_g, (c) Output signal after sampling

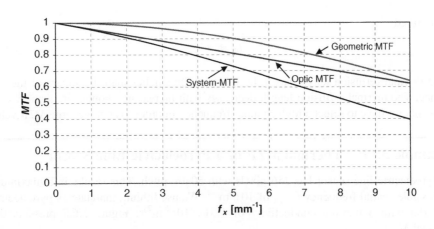

Figure 5.6.7 *MTF* for the example in Figure 5.6.6. f-number $F = 3$; wavelength $\lambda = 10\,\mu m$; pixel width $a = 50\,\mu m$

5.6.3.2 Thermal *MTF*

If heat conduction can occur between two pixels, the result will be thermal cross-talk between the signal of a pixel and the adjacent pixel. The cross-talk reduces the contrast or the modulation in the image. This is the case of pyroelectric sensors, for instance (Section 6.5). In this case, the pixels are situated close to each other on a sensor chip. Section 6.2.2 presents the thermal model of a pyroelectric sensor and the calculation of the thermal *MTF*. The thermal *MTF* for a temporal sinusoidally modulated radiation becomes

$$MTF_{th}(f_x) = \frac{1}{\sqrt{1 + \left(\dfrac{a_S\,\omega_x^2}{\omega_{Ch}}\right)^2}} = \frac{1}{\sqrt{1 + (l_{th}^2\omega_x^2)^2}} \tag{5.6.28}$$

with spatial angular frequency $\omega_x = 2\pi f_x$, chopper angular frequency $\omega_{Ch} = 2\pi f_{Ch}$, thermal conductivity a_S of the sensor element and thermal diffusion length l_{th}. This means that thermal *MTF* is a low-pass with the thermal diffusion length

$$l_{th} = \sqrt{\frac{a_S}{\omega_{Ch}}} \tag{5.6.29}$$

The largest admissible spatial frequency is the NYQUIST frequency (Equation 5.6.25). In order to achieve a maximum *MTF* at point $f_{x.Nyquist}$, it has to apply that

$$\left(\frac{\pi^2\,l_{th}^2}{r^2}\right)^2 \ll 1 \tag{5.6.30}$$

or simplified

$$l_{th}^4 \ll r^4 \tag{5.6.31}$$

The thermal diffusion length has to be substantially smaller than the square of the pixel distance. The thermal diffusion length decreases with increasing chopper frequency, which means that we have to select a high chopper frequency in order to achieve large *MTF* values.

Example 5.14:   Thermal *MTF* of a Pyroelectric Line Sensor

A pyroelectric line sensor has 256 pixels with 50 µm pitch. This results in a maximum admissible spatial frequency $f_{x,max}$ of 10 mm^{-1}. We use lithium tantalate as pyroelectric material with a thermal conductivity of 1.31×10^{-6} m^2/s. Figure 5.6.8 presents the thermal *MTF*.

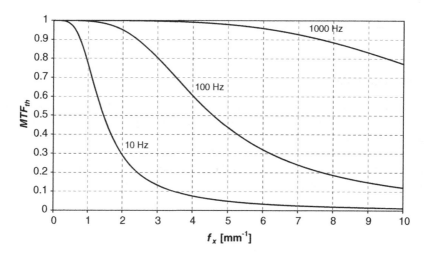

Figure 5.6.8   Thermal *MTF* of a lithium-tantalate line sensor. Parameter: chopper frequency f_{Ch}

5.6.3.3 Capacitive *MTF*

The front and back electrodes of a pyroelectric sensor constitute the capacity C_{Pixel}, where its dielectric corresponds to the pyroelectric. The input capacity C_{Gate} of the read-out electronics is connected to the pixel. The very small distances between the pixels result in capacities being generated between adjacent pixels, namely through the pyroelectric (C_{Bulk}) and the air (C_{Air}) (Figure 5.6.9). We will model a line sensor with $2n$ pixels; with n being large, for example, $n = 128$. If the middle pixel is irradiated using a point source with frequency f_{Ch}, it results in a pyroelectric current i_P with the same frequency that generates a voltage over the pixel. Due to the capacitive coupling, voltages will also be generated over the adjacent pixels. This effect is called capacitive cross-talk. This means that the cross-talk widens the input pulse and results in a decreased spatial resolution that the capacitive MTF_C describes.

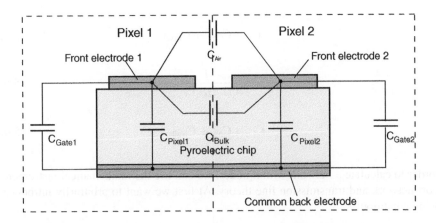

Figure 5.6.9 Capacitive coupling of two symmetrically structured pixels

For modelling the cross-talk, we can assume a symmetrical network with n two-ports. Figure 5.6.10 shows the model of a chain matrix. Here, a linear arrangement of $2n$ pixels was divided in the middle. The middle pixel ($n=0$) is excited. As it is also symmetrically divided, the exciting current corresponds to half the pyroelectric current $i_P/2$. This means that the chain matrix consists of n symmetrical π-elements. A π-element consists of two half adjacent pixel capacities Z_1 and coupling capacity Z_2. The quadripole parameters in a single π-element in Figure 5.6.10 are

$$\mathbf{K} = \begin{pmatrix} K_{11} & K_{12} \\ K_{21} & K_{22} \end{pmatrix} = \begin{pmatrix} 1 + \dfrac{Z_2}{Z_1} & Z_2 \\ \dfrac{2Z_1 + Z_2}{Z_1^2} & 1 + \dfrac{Z_2}{Z_1} \end{pmatrix} \qquad (5.6.32)$$

The impedances are calculated according to Figure 5.6.10:

$$Z_1 = \frac{2}{j2\,\pi f_{Ch} C_1} \qquad (5.6.33)$$

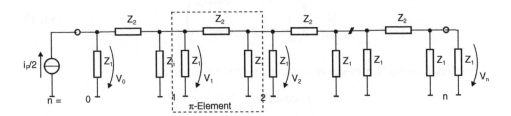

Figure 5.6.10 Chain matrix model

$$C_1 = C_{\text{Gate}} + C_{\text{Pixel}} \tag{5.6.34}$$

$$Z_2 = \frac{1}{j2\,\pi f_{\text{Ch}} C_{\text{C}}} \tag{5.6.35}$$

$$C_{\text{C}} = C_{\text{Air}} + C_{\text{Bulk}} \tag{5.6.36}$$

In order to calculate a network with n nodes, we can use the mathematical instruments of two-port network and transmission line theory. At first we want to arbitrarily introduce the following condition:

$$K_{11} = K_{22} = \cosh g \tag{5.6.37}$$

Variable g is called wave transfer factor. Due to reversibility, it applies for chain conductor matrix $\det(K) = 1$, which results in

$$\sqrt{K_{12} K_{21}} = \sinh g \tag{5.6.38}$$

With wave impedance Z_0

$$Z_0 = \sqrt{\frac{K_{12}}{K_{21}}} \tag{5.6.39}$$

it follows that

$$K_{12} = Z_0 \sinh g \tag{5.6.40}$$

and

$$K_{21} = \frac{1}{Z_0} \sinh g \tag{5.6.41}$$

The chain matrix now becomes

$$\mathbf{K} = \begin{pmatrix} \cosh g & Z_0 \sinh g \\ \dfrac{1}{Z_0} \sinh g & \cosh g \end{pmatrix} \tag{5.6.42}$$

If we link n identical π-elements, it results for the chain matrix:

$$\mathbf{K}_n = \begin{pmatrix} \cosh n\,g & Z_0 \sinh n\,g \\ \dfrac{1}{Z_0} \sinh n\,g & \cosh n\,g \end{pmatrix} \tag{5.6.43}$$

By inserting the real impedances, the wave impedance becomes

$$Z_0 = \sqrt{\frac{Z_2 Z_1^2}{2Z_1 + Z_2}} \qquad (5.6.44)$$

and the wave transfer factor results from (Equation 5.6.40):

$$g = \text{arc sinh} \frac{K_{12}}{Z_0} \qquad (5.6.45)$$

With $K_{12} = Z_2$, Equation 5.6.45 becomes

$$g = \ln\left(\frac{Z_2}{Z_0} + \sqrt{1 + \left(\frac{Z_2}{Z_0}\right)^2}\right) \qquad (5.6.46)$$

with

$$\frac{Z_2}{Z_0} = \sqrt{2\frac{Z_2}{Z_1} + \frac{Z_2^2}{Z_1^2}} \qquad (5.6.47)$$

or

$$\sqrt{1 + \left(\frac{Z_2}{Z_0}\right)^2} = \left(1 + \frac{Z_2}{Z_1}\right) \qquad (5.6.48)$$

resulting from Equation 5.6.44. Thus,

$$g = \ln\left(1 + \frac{Z_2}{Z_1} + \sqrt{2\frac{Z_2}{Z_1} + \frac{Z_2^2}{Z_1^2}}\right) = \ln\left[1 + \frac{Z_2}{Z_1}\left(1 + \sqrt{1 + 2\frac{Z_1}{Z_2}}\right)\right] \qquad (5.6.49)$$

Now we are looking for the voltage ratio of input voltage V_0 to voltage V_n after the nth element with each chain being completed after n elements with Z_0. It follows from the quadripole equation that:

$$\frac{V_n}{V_0} = \frac{1}{K_{11} + \frac{K_{12}}{Z_0}} = \frac{1}{\cosh ng + \sinh ng} = e^{-ng} \qquad (5.6.50)$$

By inserting g, we arrive at

$$\frac{V_n}{V_0} = \left(\frac{Z_2}{Z_0} + \sqrt{1 + \left(\frac{Z_2}{Z_0}\right)^2}\right)^{-n} = \left[1 + \frac{Z_2}{Z_1}\left(1 + \sqrt{1 + 2\frac{Z_1}{Z_2}}\right)\right]^{-n} \qquad (5.6.51)$$

By inserting Equations 5.6.33 and 5.6.35, the result we were looking for becomes

$$\frac{V_n}{V_0} = \left[1 + \frac{C_I}{2\,C_C} \left(1 + \sqrt{1 + 4\frac{C_C}{C_I}} \right) \right]^{-n} \qquad (5.6.52)$$

Assuming an exponential drop of the cross-talk from one pixel to the next adjacent pixel, we can define a spatial distance d according to the following equation:

$$\frac{V_n}{V_0} = e^{-nd} \qquad (5.6.53)$$

with d resulting from Equation 5.6.52:

$$d = \ln\left[1 + \frac{C_I}{2C_C} \left(1 + \sqrt{1 + 4\frac{C_C}{C_I}} \right) \right] \qquad (5.6.54)$$

Figure 5.6.11 shows the cross-talk to the adjacent pixel. It is independent of the chopper frequency and only depends on the capacity ratio C_C/C_I.

In order to calculate MTF, a signal fed in at node 0 is now interpreted as a spot radiation of the pixel. Voltage V_0 is the pixel's electrical response to the point source signal. Voltages V_n at the adjacent pixels ($n = 1, 2, 3 \ldots$) are then a cross-talk of the signal. The spatially resolved signal can be represented with DIRAC delta functions:

$$\frac{V_n}{V_0} = \sum_{n=-\infty}^{+\infty} e^{-d\,|n|}\,\delta(x - n\,r) = e^{-\frac{d}{r}|x|} \sum_{n=-\infty}^{\infty} \delta(x - n\,r) \qquad (5.6.55)$$

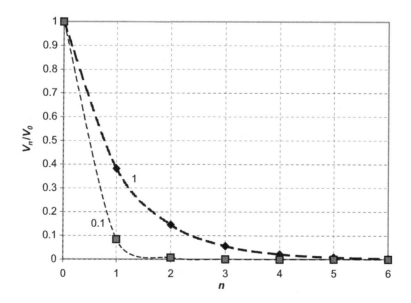

Figure 5.6.11   Cross-talk to adjacent pixels. Parameter C_C/C_I

The DIRAC delta function sum states that there is only a voltage at the pixels with distance r. In front of the sigma sign, there is the envelope curve of the pulse sequence. The pulse sequence is symmetrical to the origin. As the input signal already is a point source, *MTF* can be directly calculated from Equation 5.6.55 using a FOURIER transform. The following FOURIER correspondences apply:

$$e^{-a|x|} \Leftrightarrow \frac{2a}{a^2 + (2\pi f_x)^2} \qquad (5.6.56)$$

$$\sum_{n=-\infty}^{\infty} \delta(x-nr) \Leftrightarrow \frac{1}{r} \sum_{v=-\infty}^{\infty} \delta\left(f_x - \frac{n}{r}\right) \qquad (5.6.57)$$

Both functions are multiplied with each other in the spatial domain, that is, that they have to be convoluted in the frequency domain. A convolution with a DIRAC delta function sum means, in turn, multiplying it by itself:

$$F\left\{\frac{V_n}{V_0}\right\} = \frac{2d}{r^2} \frac{1}{\left(\frac{d}{r}\right)^2 + (2\pi f_x)^2} \sum_{n=-\infty}^{\infty} \delta\left(f_x - \frac{n}{r}\right) \qquad (5.6.58)$$

By normalising to $f_x = 0$, we now arrive at *MTF*. In the following, we will only use the envelope curve as MTF[5]:

$$MTF_C = \frac{1}{1 + \left(\frac{2\pi r}{d} f_x\right)^2} \qquad (5.6.59)$$

Figure 5.6.12 shows capacitive MTF_C. In order to arrive at a capacitive cross-talk as small as possible and thus at a capacitive MTF_C as large as possible, coupling capacity C_C has to be substantially smaller than the sum C_1 of element capacity and gate capacity.

5.6.4   Measuring the Modulation Transfer Function

There is a variety of options for measuring *MTF* that we want to present in the following [4]. Figure 5.6.13 shows the principal measuring set-up. The measuring is carried out by projecting a scene onto the sensor. In Figure 5.6.13 we have the example of a slit aperture as the scene. We will always measure the system *MTF*. To determine the sensor *MTF*, we must know the *MTF* of the optics. We always assume the sensor to be a linear system for measuring.

---

[5] The discretization is usually not included in *MTF*.

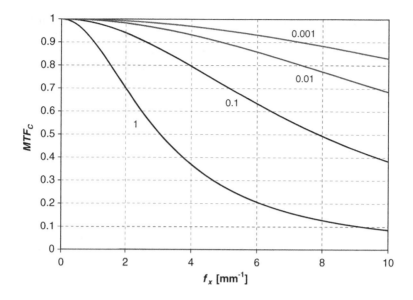

Figure 5.6.12   Capacitive *MTF*. Pixel distance $r = 50\,\mu$m; parameter: C_C/C_I

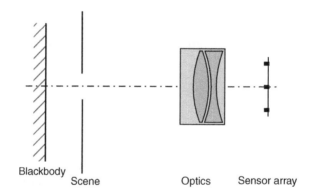

Figure 5.6.13   Measuring set-up for determining *MTF* of a sensor

5.6.4.1 Measuring with a Sinusoidal-Modulated Scene

For the projection of a sinusoidal spatial grid, we directly measure the modulation. We need to generate a spatial grid for each spatial frequency. It is rather complicated to generate a sinusoidal spatial grid for the IR range. Therefore, this method is not feasible.

5.6.4.2 Measuring with a Rectangular-Modulated Scene

When measuring with a rectangular scene, the contrast and thus *CTF* will be measured. Often *CTF* is given as *MTF* without further explanations. For high spatial frequencies both blend into each other, due to the low-pass characteristics of a sensor.

5.6.4.3 Measuring with a Point or Line-Source

A point or line source is projected to a sensor. With FOURIER transformation and normalisation, we arrive at the system's *MTF*. The point or slit width in the image has to be known, though, in order to be able to separate them out. In addition, *SNR* in the sensor is very small for small slits.

5.6.4.4 Measuring with a Knife-Edge Scene

The *MTF* can be calculated from the knife-edge image, the so-called knife-edge spread function *ESF*. The derivative of the knife-edge spread function with respect to the position results in a slit-image function. The slit-image function corresponds to the line spread function (*LSF*). This can be FOURIER-transformed and normalised in order to arrive at the *MTF*:

$$LSF(x) = \frac{\mathrm{d}\,ESF(x)}{\mathrm{d}\,x} \tag{5.6.60}$$

During the differentiation the steady component will be lost. The FOURIER transform of the line spread function results in:

$$F\{LSF(x)\} = F\left\{\frac{\mathrm{d}\,ESF(x)}{\mathrm{d}\,x}\right\} \tag{5.6.61}$$

In linear systems, FOURIER transformation and differentiation are interchangeable. Therefore, we have the option to carry out the differentiation in the frequency domain. It then applies that:

$$S(f_x) = F\{LSF(x)\} = j\,2\pi f_x\,F\{ESF(x)\} \tag{5.6.62}$$

For the FOURIER transformation of the knife-edge image we have to take into consideration that a digital FOURIER transform requires the periodic continuation of the signal. However, at the margin of the knife-edge image, there are signal interruptions, that have to be eliminated by adding a ramp, for instance. As we assume low-pass behaviour, we normalise to $f_x = 0$:

$$MTF(f_x) = \frac{S(f_x)}{S(f_x = 0)} \tag{5.6.63}$$

When differentiating in the frequency domain, we have to calculate the value at point $f_x = 0$ by extrapolating values of $S(f_x)$ with low spatial frequencies.

Example 5.15:   Measuring *MTF* Using a Knife-Edge Image

A measured signal is, for instance, the edge at $x = 3.75$ mm in Figure 5.6.6c (Figure 6.6.11). Assuming that $N = 8$ elements at the edge are considered to be a number sequence for the signal:

$$ESF(n) = \{-0.011; 0.033; -0.025; 0.312; 0.959; 0.983; 0.997; 0.992\} \tag{5.6.64}$$

The differentiation can be directly performed at the number sequence. The result is the line spread function:

$$LSF(n) = \sum_{n=1}^{7} k(n+1) - k(n) \tag{5.6.65}$$

The last element is completed with a zero:

$$LSF(n) = \{0.044;\ -0.058;\ 0.337;\ 0.647;\ 0.023;\ 0.014;\ -0.004;\ 0\} \tag{5.6.66}$$

Now we can perform a fast FOURIER transform (FFT) with these eight elements. The amplitude of FFT is

$$S(n) = \{1.003;\ 0.892;\ 0.740;\ 0.533;\ 0.203;\ 0.533;\ 0.740;\ 0.892\} \tag{5.6.67}$$

The result is symmetrical to the NYQUIST frequency at $n=4$; and therefore only values of up to $n=4$ are relevant. After normalising, it results the following MTF:

$$MTF(n) = \{1;\ 0.889;\ 0.738;\ 0.531;\ 0.203\} \tag{5.6.68}$$

The spatial frequencies of the MTF values can be calculated according to the following formula:

$$f_x(n) = \frac{n}{Nr} \tag{5.6.69}$$

For a pixel pitch of $r=50\,\mu$m, it applies that:

$$f_x(n) = \{0;\ 2.5;\ 5.0;\ 7.5;\ 10\} \tag{5.6.70}$$

with the unit for all values being mm^{-1}. In order to arrive at sensor MTF, we eventually have to divide by the optical MTF in Figure 5.6.2:

$$MTF_{Optic}(n) = \{1;\ 0.968;\ 0.936;\ 0.905; 0.873\} \tag{5.6.71}$$

Sensor MTF then is

$$MTF_{Sensor}(n) = \{1;\ 0.918;\ 0.788;\ 0.587;\ 0.232\} \tag{5.6.72}$$

Figures 5.6.14 and 5.6.15 provide a graphic representation of the results. In order to achieve a higher resolution, we use $N=32$ for the graphic representation.

For the following reasons, there are deviations between the given sensor MTF in Figure 5.6.7 and the calculated value:

- For frequencies that are higher than the NYQUIST frequency, MTF is not zero in order to cause a cross-talk (aliasing). Values close to the NYQUIST frequency of 10 mm^{-1} can be manipulated by changing the position of the edge in relation to the pixels.

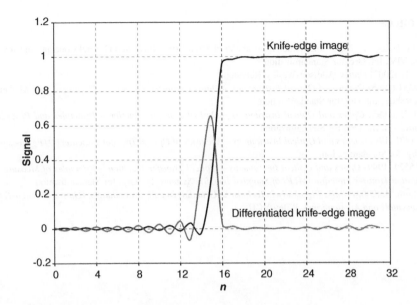

Figure 5.6.14   Knife-edge image (section of the signal curve in Figure 5.6.6c, point $n = 16$ is located at 3.75 mm) and differentiated knife-edge image for calculating *MTF*

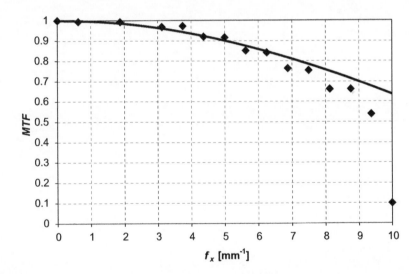

Figure 5.6.15   Measured sensor *MTF*. Full line: *MTF* according to Figure 5.6.7 Dots 'measured' values

- The fluctuations of the input signal also appear in the image range.
- We have used a very small width of the edge image function ($N = 32$). For the calculated example with $N = 8$ pixels, the measuring error is inadmissibly high!
- There are rounding errors in the calculation.

References

1. ISO/IEC Guide 99-12:2007 International Vocabular of Metrology – Basic and General Concepts and Associated Terms, VIM, Int. Org. for Standardization.
2. Hecht, E. (2003) *Optics*, Addison-Wesley, Reading.
3. ISO 9334 (1995) *Optics and Optical Instruments – Optical Transfer Function – Definitions and* Mathematical Relationships, Int. Org. for Standardization.
4. ISO 9335 (1995) *Optics and Optical Instruments – Optical Transfer Function – Principles and Procedures of Measurement,* Int. Org. for Standardization.
5. ISO 11421 (1997) *Optics and Optical Instruments – Accuracy of Optical Transfer Function (OTF) Measurement,* Int. Org. for Standardization.
6. ISO 15529 (1999) *Optics and Optical Instruments – Optical Transfer Function – Principles of Measurement of Modulation Transfer Function (MTF) of Sampled Imaging Systems,* Int. Org. for Standardization.
7. Haferkorn, H. (2003) *Optik: Physikalisch-Technische Grundlagen und Anwendungen (Optics: Physico-Technical Fundamentals and Applications)*, Wiley-VCH.

6

Thermal Infrared Sensors

Thermal infrared sensors are radiation detectors that experience a change in temperature due to absorption of infrared radiation and convert this temperature change into an electric output signal. Thermal infrared sensors are also called radiation temperature sensors. Their function principle is thus different from that of semiconductor-based photon or quantum sensors where – due to different photoelectric effects – the photons of the radiation generate charge carriers. In order to be able to detect the low-energy infrared radiation, photon sensors have to be cooled substantially below ambient temperature. As opposed to this, thermal infrared sensors can be operated at ambient temperature. The radiation noise that limits temperature resolution in thermal sensors has a $\sqrt{T}$-dependence, which means that cooling does not substantially improve detectivity. Therefore, they are particularly suitable for small, light and portable applications. The use of modern microelectronic and micromechanical manufacturing methods has resulted in a rapid development of miniaturised and high-resolution sensors that are inexpensive and open up an ever increasing number of new application areas, such as thermography.

6.1 Operating Principles

Thermal radiation sensors convert radiant flux Φ_S into an electric signal (voltage V_S or current I_S). Figure 6.1.1 represents the measuring chain.

Although they use a whole variety of different physical operating principles, the basic structure of all thermal IR sensors is identical (Figure 6.1.2). The incident IR radiant flux is absorbed by a thermally well-isolated detector element (pixel) and converted into heat. The temperature of the pixel is directly proportional to the power of the absorbed IR radiation. Therefore, the responsivity of a thermal sensor in principle is wavelength-independent. However, in many cases there is a wavelength dependence of responsivity due to the absorption characteristics of the pixel. If needed, the IR radiation can be temporally modulated with a chopper.

The conversion of the pixel temperature into an electrical signal depends on the sensor type. We distinguish between energy converters that utilise thermoelectric transducer effects and parametric transducers where temperature modulates an electric signal [1].

Thermal Infrared Sensors: Theory, Optimisation and Practice   Helmut Budzier and Gerald Gerlach
© 2011 John Wiley & Sons, Ltd

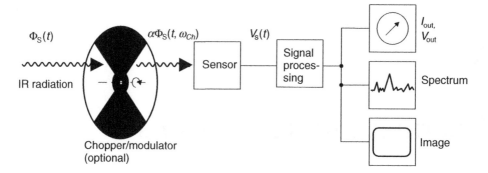

Figure 6.1.1 Measuring chain for determining infrared radiation

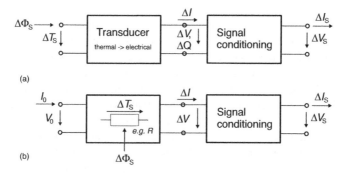

Figure 6.1.2 Mode of operation of thermal sensors (a) with thermoelectric energy conversion and (b) with electric signal modulation in parametric transducer

Thermal IR sensors that operate according to the energy transducer principle apply the following physical principles:

- SEEBECK effect: thermocouple elements or thermopiles,
- pyroelectric effect: pyroelectric sensors.

Parametric sensors use the temperature to modulate the relation between electric supply parameter and electric output parameter, that is an electric energy conversion. They need an auxiliary power source (electric supply energy). We utilise, amongst others, the following effects:

- dependence of the electric resistance on the temperature (resistor bolometer[1]).
- dependence of the pressure in a closed volume on the temperature (GOLAY cell),

---

[1] Bolometer means in general radiation meter (greek: *bolo* = ray). However, it has become general usage that only resistance-dependent radiation temperature sensors are called bolometers.

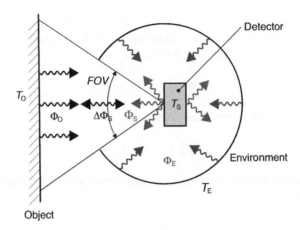

Figure 6.1.3 Radiation exchange at an infrared sensor

- dependence of the mechanical tension on the temperature,
- dependence of the forward voltage of a diode on the temperature.

A thermal IR sensor usually consists of a thermally well-isolated very thin chip, which is normally called detector element. The temperature of the detector element is determined by the absorbed radiant flux Φ_A and the radiant flux Φ_S emitted by the detector element. Figure 6.1.3 shows the radiation exchange between detector element, object and detector surroundings (background). We do not take into account a possible heat flow between detector and its surroundings through thermal conduction via the sensor mounting, for instance (see Section 6.2). The detector is situated in a vacuum.

The absorbed radiant flux Φ_A consists of object radiation Φ_O and environment radiation Φ_E:

$$\Phi_A = \bar{\alpha}(\Phi_O + \Phi_E) \qquad (6.1.1)$$

with band absorption $\bar{\alpha}$. The emitted radiant flux Φ_S is irradiated into the entire half-space:

$$\Phi_S = \bar{\varepsilon}\, \sigma\, T_S^4\, A_S \qquad (6.1.2)$$

with band emission $\bar{\varepsilon}$ and sensor area A_S. Here we will not take into account possible emission und absorption at the back of the detector element. It is possible to include the backside emission and absorption in the calculation by doubling the sensor area. For radiation exchange $\Delta\Phi_{A\text{-}S}$ at the sensor, it now applies that

$$\Delta\Phi_{A\text{-}S} = \bar{\alpha}(\Phi_O + \Phi_E) - \bar{\varepsilon}\, \Phi_S \qquad (6.1.3)$$

The starting point of the following considerations is a hypothetical equilibrium where the environment and the object are supposed to have the same temperature ($T_O = T_E$). Due to energetic reasons, the sensor thus has to have the same ambient temperature ($T_S = T_E$). This state is called operating point. The operating point temperature is also called background temperature T_B ($T_B = T_E$). It is of course also possible to have other operating points with $T_O \neq T_E$; this results in the following equations in only one additional constant, though. The sensor receives radiation from the entire half-space:

$$\Phi_A = \bar{\alpha}\,\sigma\,T_E^4\,A_S \tag{6.1.4}$$

We reach a thermal equilibrium when $\Delta\Phi_{A\text{-}S} = 0$:

$$\bar{\alpha}\,\sigma\,T_E^4\,A_S = \bar{\varepsilon}\,\sigma\,T_S^4\,A_S \tag{6.1.5}$$

Assuming that sensor temperature T_S and ambient temperature T_E are equal in the operating point, it follows that

$$\bar{\alpha} = \bar{\varepsilon} \tag{6.1.6}$$

This means that in thermal equilibrium, the band absorption and band emission are equal. This does not apply to the specific wavelengths, though.

Changing object temperature T_O results in a modulation of the sensors above the defined operating point T_S. This radiant flux change $\Delta\Phi_S$ due to the object temperature change is

$$\Delta\Phi_S = \Phi_O(T_O) - \Phi_O(T_E) \tag{6.1.7}$$

Thus Equation 6.1.1 becomes

$$\Phi_A = \bar{\alpha}[\Phi_O(T_E) + \Delta\Phi_S(T_O) + \Phi_E(T_E)] = \bar{\alpha}[\Phi_B(T_E) + \Delta\Phi_S(T_O)] \tag{6.1.8}$$

The sum of background radiation $\Phi_B = \Phi_O(T_E) + \Phi_E(T_E)$ and radiant flux change $\Delta\Phi_S$ of the object is the absorbed radiant flux Φ_A. This means that the signal the sensor receives is background radiation Φ_B from the whole half-space and in addition radiant flux $\Delta\Phi_S$ from the *FOV*. Radiant flux $\Delta\Phi_S$ corresponds to the radiant flux exchanged between sensor and object and is calculated according to Equations 3.2.22, 3.2.32 and 3.2.41 as

$$\Delta\Phi_S = \sigma(T_O^4 - T_E^4)A_S \sin^2\frac{FOV}{2} \tag{6.1.9}$$

Now we will look at a small modulation around operating point T_S. Applying Equation 6.1.3 and inserting Equations 6.1.2 and 6.1.4, it results that

$$\Delta\Phi_{A\text{-}S} = \Phi_A - \Phi_S = \bar{\alpha}\,\Phi_O + \bar{\alpha}\,\Phi_E - \bar{\varepsilon}\,\Phi_S = \bar{\alpha}\,\Phi_O - \bar{\varepsilon}\,A_S\,\sigma\left(T_S^4 - \frac{\bar{\alpha}}{\bar{\varepsilon}}T_E^4\right) \tag{6.1.10}$$

Applying the condition in Equation 6.1.6, the above equation becomes

$$\Delta\Phi_{A\text{-}S} = \bar{\alpha}\,\Phi_O - \bar{\varepsilon}\,A_S\,\sigma(T_S^4 - T_E^4) \tag{6.1.11}$$

For the thermal equilibrium it applies due to $\Delta\Phi_{A\text{-}S} = 0$ that

$$\bar{\alpha}\,\Phi_O = \bar{\varepsilon}\,A_S\,\sigma(T_S^4 - T_E^4) \tag{6.1.12}$$

Deriving Equation 6.1.12 for a constant ambient temperature T_E in regards to temperature T_S, it results that

$$\bar{\alpha}\frac{d\Phi_O}{dT_S} = 4\,\bar{\varepsilon}\,A_S\,\sigma\,T_S^3 \qquad (6.1.13)$$

Equation 6.1.13 describes the slope of the characteristic curve of radiant flux and temperature in point T_S. This is by definition a thermal conductance, in this case through radiation:

$$G_{th,S} = 4\bar{\varepsilon}A_S\sigma T_S^3 \qquad (6.1.14)$$

For the transition to the difference quotient $(d\Phi_O \rightarrow \Delta\Phi_S;\ dT_S \rightarrow \Delta T_S)$, Equation 6.1.13 becomes

$$\bar{\alpha}\,\Delta\Phi_S = 4\bar{\varepsilon}A_S\sigma T_S^3\Delta T_S \qquad (6.1.15)$$

With Equation 6.1.14, it now applies that

$$\bar{\alpha}\,\Delta\Phi_S = G_{th,S}\Delta T_S \qquad (6.1.16)$$

This results in a temperature change in the sensor that is caused by the changed object radiation

$$\Delta T_S = \frac{\bar{\alpha}\,\Delta\Phi_S}{G_{th,S}} \qquad (6.1.17)$$

The conversion of temperature change ΔT_S into signal voltage ΔV_S is sensor-specific. In the operating point, the sensor element has temperature T_E and the respective bias output signal I_0 or V_0. If now object temperature T_O changes, Equation 6.1.9 applies to the exchanged radiant flux, and there will be a temperature change ΔT_S in the sensor. Consequently, the sensor output signal changes by ΔV_S or ΔI_S.

If we temporally modulate the object radiation, we will arrive at mean sensor temperature $\overline{T_S(t)}$. Assuming that the modulator (chopper) has ambient temperature T_E the mean operating point temperature becomes

$$\overline{T_S(t)} = T_E + \frac{\bar{\alpha}}{G_{th}}\overline{\Delta\Phi_S(t)} \qquad (6.1.18)$$

We will always assume that object temperature T_O is a fixed given value. There is no feedback to the measuring object. As optical projections are always reversible, not only is the measuring object projected onto the sensor, but also the sensor is projected onto the measuring object. This can cause a change in the radiation condition at the object, however, and consequently lead to a change in object temperature. In practice, however, this effect can be neglected in almost all cases. Exceptions are thermally well-isolated measuring objects, such as thermal IR sensors.

Example 6.1:   Temperature Change of a Sensor

We want to calculate temperature change ΔT_S as a function of the object temperature T_O for the sensor arrangement presented in Figure 6.1.3. Operating point temperature is $\vartheta_S = \vartheta_E = 23\,°C$, field of view $FOV = 60°$, mean emissivity $\bar{\varepsilon} = 1$ and sensor area $A_S = 1\,mm^2$. The thermal conductance due to radiation is calculated according to Equation 6.1.14:

$$G_{th,S} = 5.89 \times 10^{-6}\ K/W$$

Applying Equations 6.1.9, 6.1.14 and 6.1.17 with $\bar{\varepsilon} = \bar{\alpha}$, the temperature change we are looking for becomes

$$\Delta T_S = \frac{\bar{\alpha}\,\Delta\Phi_S}{G_{th,S}} = \frac{\bar{\alpha}\,\sigma(T_O^4 - T_S^4)\,A_S\,\sin^2\dfrac{FOV}{2}}{4\,\bar{\varepsilon}\,A_S\,\sigma\,T_S^3} = \frac{1}{4}\left(\frac{T_O^4}{T_S^3} - T_S\right)\sin^2\frac{FOV}{2} \qquad (6.1.19)$$

We have to take into account that this equation applies to only small modulation ΔT_S. Figure 6.1.4 shows the calculated temperature difference. The presented values are the theoretically possible temperature changes in the sensor. In practice, heat conduction due to sensor mountings and surrounding gas has to be taken into account.

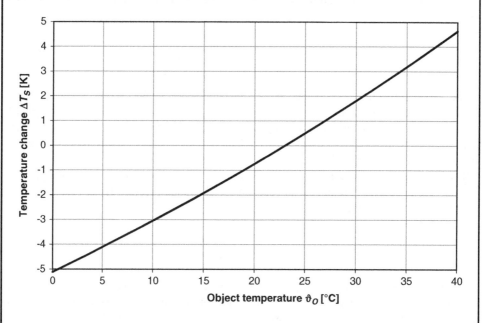

Figure 6.1.4 Temperature change in the sensor according to Figure 6.1.3. $FOV = 60°$; $\vartheta_S = \vartheta_U = 23\,°C$

6.2 Thermal Models

Calculating temperature change ΔT_S of the detector element due to the absorption of a radiant flux $\Delta\Phi_S$ is the key problem of the theoretical analysis and simulation of thermal sensors. In the following, we will present some alternatives for calculating thermal conditions. The starting point is the sensor structure provided in Figure 6.1.3. The sensor element is always a thin chip that is heated by a radiant flux and interacts with its environment by heat conduction.

6.2.1 Simple Thermal Model

According to Figure 6.2.1, we assume that the absorption of the radiant flux causes a homogeneous temperature distribution in the sensor. The heat energy Q of the incident radiant flux $\Delta\Phi_S$ heats the sensor element with its heat capacity C_{th}. Here a cuboid detector element (thin chip) with area A_S and thickness d_S has the heat capacity

$$C_{th} = c'_S \rho_S A_S d_S \qquad (6.2.1)$$

with specific heat capacity c'_S and mass density ρ_S. At the same time, heat energy is dissipated via several heat conduction mechanisms to the environment. The heat conduction to the surroundings can be represented as thermal resistance R_{th} or thermal conductance $G_{th} = 1/R_{th}$.

The energy balance or power balance thus results in

$$\alpha\,\Delta\Phi_S(t) = P_C(t) + P_G(t) \qquad (6.2.2)$$

with power flow P_C through the heat capacity of the sensor element and power flow P_G through the thermal resistances to the thermal ground. With $P_C = dQ_C/dt$ and Equation 6.2.2, it follows for the time domain that

$$\alpha\,\Delta\Phi_S(t) = C_{th}\frac{d[\Delta T_S(t)]}{dt} + G_{th}\Delta T_S(t) \qquad (6.2.3)$$

The time dependence of the absorbed radiant flux $\Delta\Phi_S$ that is incident on the sensor is, in principle, arbitrary. For a temporally constant radiant flux, it applies that $\alpha\Delta\Phi_S(t) = \text{const.}$

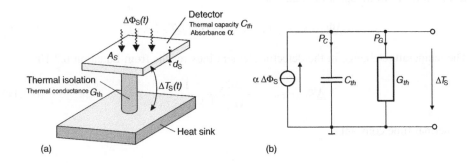

Figure 6.2.1   Simple model of a thermal IR sensor. (a) Principle; (b) thermal network model

In this case, it results from Equation 6.2.3 for the temperature change of the detector element (see also Section 6.1):

$$\Delta T_S = \frac{\alpha \, \Delta \Phi_S}{G_{th}} \tag{6.2.4}$$

If we now modulate the amplitude of the radiant flux with a chopper, there results a periodic time function of the radiant flux. Here we assume a constant amplitude ($\alpha \Delta \Phi_S = \text{const}$). These periodic time functions can be mathematically expressed by sinusoidal processes, such as FOURIER series. Harmonic analysis leads itself particularly to this calculation, and it follows from Equation 6.2.3 for frequency ω that

$$\alpha \, \underline{\Delta \Phi}_S = (j\omega C_{th} + G_{th}) \, \underline{\Delta T}_S \tag{6.2.5}$$

In the frequency or image domain, respectively, we can use LAPLACE transform to simply express any time function by complex frequencies $s = \sigma + j\omega$. It results from Equation 6.2.3

$$\alpha \underline{\Delta \Phi}_S = (\underline{s} C_{th} + G_{th}) \underline{\Delta T}_S \tag{6.2.6}$$

6.2.1.1 Temperature Change for Harmonic Excitation in the Frequency Domain

For a harmonic excitation, radiant flux $\Delta \Phi(t)$ has a sinusoidal time function with modulation or chopper frequency $\omega_{Ch} = 2\pi f_{Ch}$ and amplitude $\Delta \hat{\Phi}_S$:

$$\Delta \Phi_S(t) = \Delta \hat{\Phi}_S \cos \omega_{Ch} t \tag{6.2.7}$$

It then applies in frequency domain that

$$\underline{\Delta \Phi}_S(\omega_{Ch}) = \Delta \Phi_S \, e^{j\omega_{Ch} t} \tag{6.2.8}$$

with effective value

$$\Delta \Phi_S = \frac{\Delta \hat{\Phi}_S}{\sqrt{2}} \tag{6.2.9}$$

It now follows from Equation 6.2.3 that

$$\alpha \, \underline{\Delta \Phi}_S(\omega_{Ch}) = (j\omega_{Ch} C_{th} + G_{th}) \, \underline{\Delta T}_S(\omega_{Ch}) \tag{6.2.10}$$

The temperature change in the detector element thus results from Equation 6.2.10:

$$\underline{\Delta T}_S(\omega_{Ch}) = \frac{\alpha \, \underline{\Delta \Phi}_S}{G_{th}} \frac{1}{1 + j\omega_{Ch} \tau_{th}} \tag{6.2.11}$$

with thermal time constant

$$\tau_{th} = \frac{C_{th}}{G_{th}} \tag{6.2.12}$$

Only the effective value of the temperature change, that is the absolute value,[2] is of interest:

$$\left|\underline{\Delta T}_S(\omega_{Ch})\right| = \Delta T_S(\omega_{Ch}) = \frac{\alpha\,\Delta\Phi_S}{G_{th}}\frac{1}{\sqrt{1+\omega_{Ch}^2\tau_{th}^2}} \qquad (6.2.13)$$

That means that a detector element behaves like a first-order low-pass. It applies to unmodulated radiant flux, that is $\omega_{Ch}=0$, that:

$$\Delta T_S(\omega_{Ch}=0) = \frac{\alpha\,\Delta\Phi_S}{G_{th}} \qquad (6.2.14)$$

This corresponds to the above described case of temporally constant radiant flux (Equation 6.2.4). In order to evaluate the thermal conditions in a detector element we usually normalise the temperature change to the maximum temperature change. We arrive at this value using Equation 6.2.14 for constant-light-sensitive sensors if the thermal conductance coefficient is only given by radiation and the absorption $\alpha=1$:

$$\Delta T_{S\,max} = \frac{\Delta\Phi_S}{G_{th,S}} \qquad (6.2.15)$$

This results in the so-called normalised temperature responsivity T_R:

$$T_R = \frac{\left|\underline{\Delta T}_S(\omega_{Ch}=0)\right|}{\Delta T_{S\,max}} = \frac{\alpha G_{th,S}}{G_{th}} \qquad (6.2.16)$$

For alternating-light-sensitive sensors, such as pyroelectric sensors, we usually normalise to Equation 6.2.13, and it applies that

$$T_R = \frac{\left|\Delta T_S\omega_{Ch}\right|}{\left|\Delta T_{S\,max}(\omega_{Ch})\right|} \qquad (6.2.17)$$

The maximum of normalised temperature responsivity T_R is 1.

Example 6.2: Normalised Temperature Responsivity of a Microbolometer Bridge

A microbolometer bridge is a micromechanical bridge that is located approximately 2.5 µm above a Si circuit. We want to analyse the effect that thermal conductance $G_{th,Gas}$ through the gas layer (dry nitrogen gas) between pixel and circuit has on normalised temperature responsivity T_R. In addition to the heat flux due to radiation, heat dissipates via the mounting, the microbridge pillars and the gas surrounding the microbridge. The thermal conductance is the sum of the thermal conductance due to radiation $G_{th,R}$, the thermal conductance through the mechanical support of the pixel $G_{th,Leg}$ and through the thermal conductance of gas $G_{th,Gas}$:

---

[2] The phase of the temperature change causes a temporal offset between the amplitudes of the radiant flux change and the temperature change; it is often irrelevant for a further signal analysis.

$$G_{th} = G_{th,S} + G_{th,Leg} + G_{th,Gas} \tag{6.2.18}$$

Thus the normalised temperature responsivity T_R becomes:

$$T_R = \frac{G_{th,S}}{G_{th,S} + G_{th,Leg} + G_{th,Gas}} \tag{6.2.19}$$

The thermal conductance through the bridge mounting is given with $G_{th,Leg} = 10^{-7}\,\text{W/K}$. The thermal conductance through the air can be calculated using the following dimensioning equation:

$$G_{th} = \frac{\lambda_{Gas}}{h} A_S \tag{6.2.20}$$

with thermal conductivity of the gas λ_{Gas}, height of the microbridge $h = 2.5\,\mu m$ and sensor area A_S. Here we neglect the heat flux from the sensor element to the far-away sensor window.

For a given temperature T and a large volume, heat conductivity λ_{Gas} of an ideal gas is independent of pressure p and thus of the density of the gas. Only for a mean free path of the gas molecule in the order of the dimensions of the device, heat conduction depends on pressure. For a normal pressure of 101.3 kPa (760 Torr), the mean free path is approximately 10^{-7} m. The heat conductivity of a gas can now be calculated for a constant temperature using the following equation [2]:

$$\frac{1}{\lambda_{Gas}} = \frac{1}{\lambda_{HP}} + \frac{1}{\lambda_{LP}} \tag{6.2.21}$$

Here, λ_{HP} is the portion of the heat conductivity that is independent of pressure and predominant during high pressure:

$$\lambda_{HP} = \frac{2}{3\sqrt{\pi}} \frac{c'}{\sigma_0} \sqrt{\frac{M}{N_A} k_B T} \tag{6.2.22}$$

with cross-section σ_0 of the gas molecules ($\sigma_0 = \pi d^2$), molecule diameter d, specific heat c', AVOGADRO constant N_A and molar mass M. With the parameters given in Table 6.2.1, we arrive at a value of 0.026 W/(K m) for λ_{HP}), which corresponds to the common tabulated value for nitrogen at normal pressure. The parameter λ_{LP} describes the dependence of the heat conductance coefficient of the gas on pressure p and device dimension h (dominating at low pressure):

$$\lambda_{LP} = 3c'h\,p\sqrt{\frac{8}{\pi}\frac{M}{RT}} \tag{6.2.23}$$

with universal gas constant R. This means that heat conduction is proportional to molar mass M and becomes substantially smaller when we use heavy gases, such as xenon ($M = 131\,\text{kg/mol}$). Figure 6.2.2 represents the thermal conductivity and Figure 6.2.3 the thermal conductance as a function of pressure.

Table 6.2.1   Values of the applied parameters and constants

Denomination	Symbol	Value
Height of the bridge	H	$2.5\,\mu m$
Area of the sensor element	A_S	$25 \times 25\,\mu m^2$
Cross-section	σ_o	$2.12 \times 10^{-19}\,m^2$
Specific heat of nitrogen	c'	$1040\,J/(kg\;K)$
Molar mass of nitrogen	M	$28\,kg/mol$
Universal gas constant	R	$8314\,J/(K\;mol)$
AVOGADRO constant	N_A	$6.022 \times 10^{23}\,mol^{-1}$

The calculated values can be used to determine the normalised temperature responsivity. With a thermal conductance coefficient through irradiation at $T = 300\,K$ of

$$G_{th,S} = 4\,\sigma\,T^3\,A_S = 3.83 \times 10^{-9}\,W/K$$

the normalised temperature responsivity is presented in Figure 6.2.4.

The maximum temperature responsivity is reached at maximum pressures of $10\,Pa$ (0.1 mbar). However, it amounts to only 1/30 of the maximum value of 1. The reason is the high heat dissipation through the pillars of the microbridge. If the dissipation is reduced pressure has to decrease accordingly in order to maximise responsivity.

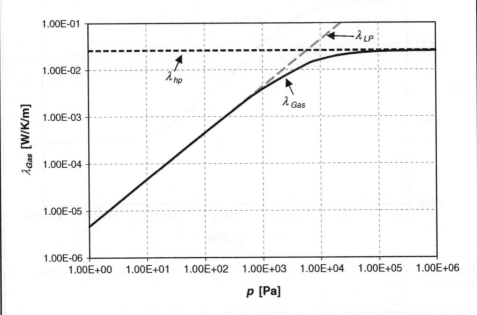

Figure 6.2.2   Thermal conductivity λ_{Gas} versus gas pressure p. Parameters: see Table 6.2.1

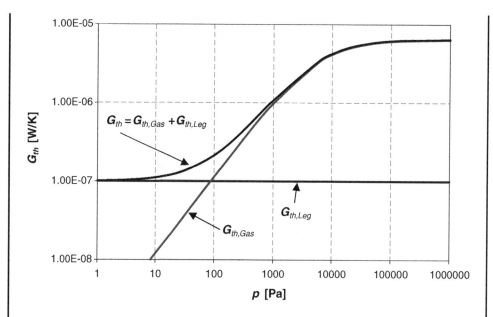

Figure 6.2.3   Thermal conductance G_{th} versus gas pressure p. Parameters: see Table 6.2.1

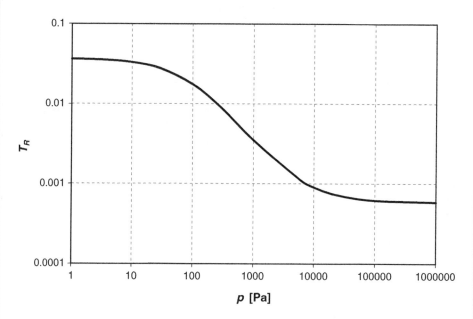

Figure 6.2.4   Normalised temperature responsivity T_R in relation to gas pressure p. Parameters: see Table 6.2.1

6.2.1.2 Temperature Change for Step Excitation

For a step change of the incident radiant flux:

$$\Delta\Phi_S(t) = 0 \qquad \text{for} \quad t \le 0$$
$$\Delta\Phi_S(t) = \Delta\Phi_{Son} \quad \text{for} \quad t > 0$$

we have to apply complex frequencies s instead of harmonic frequencies $j\omega$. Equation 6.2.11 thus becomes

$$\underline{\Delta T}_S(\underline{s}) = \frac{\alpha}{G_{th}} \frac{\underline{\Delta\Phi}_S(\underline{s})}{1 + s\,\tau_{th}} \frac{1}{} \tag{6.2.24}$$

According to the LAPLACE transform, it results for the step function that

$$\underline{\Delta\Phi}_S(\underline{s}) = \frac{\Delta\Phi_{Son}}{s} \tag{6.2.25}$$

It follows that

$$\underline{\Delta T}_S(\underline{s}) = \frac{\alpha}{G_{th}} \frac{1}{s + s^2 \tau_{th}} \Delta\Phi_{Son} \tag{6.2.26}$$

and the LAPLACE inverse transform results in

$$\Delta T_S(t) = \frac{\alpha\,\Delta\Phi_{Son}}{G_{th}} \left(1 - e^{\frac{t}{\tau_{th}}}\right) \tag{6.2.27}$$

For $t \to \infty$, the steady final value becomes

$$\Delta T_{Son} = \frac{\alpha\,\Delta\Phi_{Son}}{G_{th}} \tag{6.2.28}$$

(see Equation 6.1.4).

If we interrupt radiant flux $\Delta\Phi_{Son}$ at time $t = t_0$, it results with the following conditions

$$\Delta\Phi_S(t) = \Delta\Phi_{Son} \quad \text{for} \quad t \le t_0$$
$$\Delta\Phi_S(t) = 0 \qquad \text{for} \quad t > t_0$$

analogously

$$\Delta T_S(t \ge t_0) = \Delta T_{Son} e^{-\frac{t - t_0}{\tau_{th}}} \tag{6.2.29}$$

For $t \to \infty$, we arrive at steady final value $\Delta T_{Soff} = 0$. Figure 6.2.5 presents the time functions for step radiant flux changes.

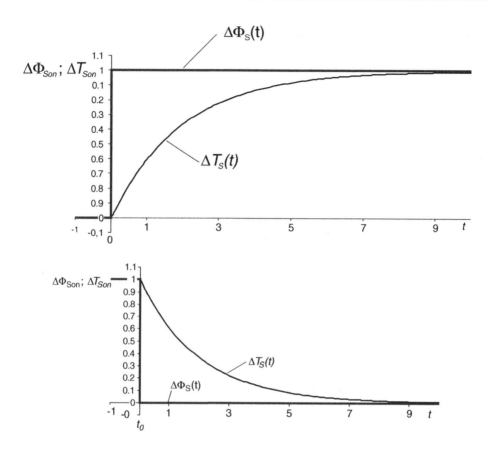

Figure 6.2.5 Temperature change for radiant flux step. (a) Switching on; (b) switching off

---

Example 6.3: Rectangular Modulation of the Radiant Flux

A rectangular modulation of the radiant flux can be interpreted as an infinite sequence of turning on and off radiant flux $\Delta\Phi_S$. Figure 6.2.6 presents the temperature change in the sensor element. The signal form of the temperature change depends heavily on the thermal time constant. The transient conditions occurring during the change of the radiant flux have not been included. We have to take into consideration that the mean temperature of the detector element $\overline{T}_S$ does not correspond to ambient temperature T_E any longer (Equation 6.1.18). In the calculated example, the mean temperature is independent of the time constant ($T_S = T_E$ for $\Delta T_S = 0$):

$$\overline{T}_S = T_E + \frac{1}{2}\Delta T_{S0} \tag{6.2.30}$$

In Figure 6.2.6, the mean values are shown as dashed lines.

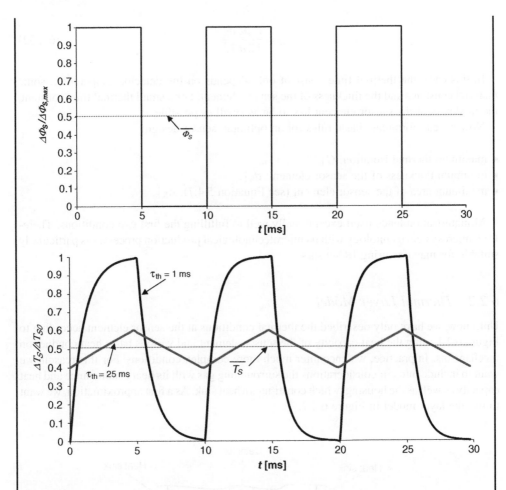

Figure 6.2.6 Time function of the temperature change for a rectangular chopping of the radiant flux. (a) Input flux difference $\Delta\Phi_S$ with a chopper frequency of 100 Hz. (b) Temperature change ΔT_S at different time constants. Dashed line: mean values

In order to achieve the largest temperature change possible, there has to be a maximum thermal isolation of the detector element, that is $G_{th} \rightarrow 0$. However, the thermal conductance through radiation (Equation 6.1.14) sets a physical limit. Therefore, additional thermal conductance $G_{th,B}$, that is caused by the mounting of the sensor element, has to fulfil the following condition:

$$G_{th,B} \ll G_{th,S} \qquad (6.2.31)$$

The thermal time constant determines the reaction time of the sensor and therefore has to be as small as possible. Assuming that the thermal conductance through radiation is the determining one, the thermal time constant can be calculated using Equations 6.2.1, 6.2.12 and 6.1.14 as follows:

$$\tau_{th} = \frac{c' \rho_S d_S}{4 \bar{\varepsilon} \sigma T_S^3}$$ (6.2.32)

In this case, the thermal time constant only depends on the detector temperature, some material constants and the thickness of the sensor element. For a small thermal time constant, the thickness of the sensor element has to be as small as possible.

Now we can formulate basic rules for an optimum sensor design:

- maximum thermal isolation: $G_{th}\downarrow$,
- minimum thickness of the sensor element: $d_S\downarrow$,
- maximum area of the sensor element (see Equation 5.4.7): $A_S\uparrow$.

Miniaturisation lends itself exceptionally well to fulfilling the first two conditions. Therefore, microsystem technology with its microtechnological production processes is particularly suitable for manufacturing IR sensors.

6.2.2 Thermal Layer Model

Until now, we have only described the thermal conditions at the sensor element according to Figure 6.2.1 using the heat capacity of the sensor element and various basic heat conduction mechanisms. In practice, we encounter much more complex conditions. For this reason, we want to include into our considerations the surrounding gas with its heat conductance and heat capacity as well as the housing, which constitutes a heat sink. As a first approximation, we want to use the layer model in Figure 6.2.7.

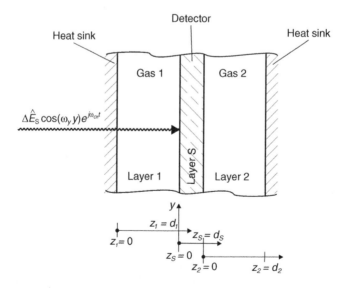

Figure 6.2.7   Model of a self-supporting detector chip. In order to simplify calculation, each layer will have its own coordinate system

As opposed to the simple thermal model, in the layer model the incident radiant flux has – in addition to the temporal dependence (ω_{Ch}) – a spatially sinusoidal distribution with spatial frequency $\omega_y = 2\,\pi\,f_y$:

$$\underline{\Delta E}_S(y, \omega_{Ch}) = \Delta \hat{E}_S \cos(\omega_y y) e^{j\omega_{Ch} t} \qquad (6.2.33)$$

This way the layer model makes it possible to calculate the thermal modulation transfer function MTF_{th} of a thermal sensor. We will present a model consisting of three layers. One layer is the sensor element (layer S) and the other two layers (layers 1 and 2) lie in between the sensor element and the thermal ground (Figure 6.2.7). We assume the layers to have an infinite expansion in the xy-plane. If layers 1 and 2 are gases (e.g. air, dry nitrogen or vacuum), we talk about self-supporting detector chips.

For calculating temperature change $\Delta T_S(\omega_{Ch}, \omega_y)$ in the detector element, we have to take into consideration the temperature change in the adjacent layers. The heat exchange of the sensor element with its environment only occurs through heat radiation and through heat conduction via these layers. The detector element is characterised by its thermal diffusivity a_S and its thermal conductivity λ_S; and layers 1 and 2 are characterised by their thermal diffusivities a_1 and a_2 as well as their thermal conductivities λ_1 and λ_2. Thermal diffusivity a and thermal conductivity λ are coupled to each other via heat capacity c' and density ρ:

$$a = \frac{\lambda}{c'\rho} \qquad (6.2.34)$$

The thermal diffusivity thus includes heat capacity C_{th} (Equation 6.2.1). The thickness of the sensor element is d_S. The thicknesses of the layers are d_1 and d_2, correspondingly. Sensor area A_S is perpendicular to edge lengths a in the x-direction and b in the y-direction. The sensor area is situated symmetrically at the coordinate origin.

We can use the heat transfer equation to accurately calculate temperature change $\Delta T_i(x, y, z_i, t)$ in layer i ($i = \in \{1, 2, S\}$):

$$\nabla^2 \Delta T_i(x, y, z_i, t) - \frac{1}{a_i} \frac{\partial \Delta T_i(x, y, z_i, t)}{\partial T_i} + \frac{q_i'}{\lambda_i} = 0 \qquad (6.2.35)$$

with LAPLACE operator ∇^2, thermal diffusivity a_i, thermal conductivity λ_i and heat power q_i' generated per volume element as the source term. In order to achieve a closed solution, we use the following simplifications:

- No heat is generated in the layers: $q_i' = 0$.
- We look at steady-state sinusoidal processes with the absorbed radiant flux being defined according to Equation 6.2.33. Radiant flux density $\underline{\Delta E}_S$ is modulated temporally sinusoidally using angular frequency ω_{Ch} and, in the y-direction, spatially cosinusoidally using spatial angular frequency ω_y.

In the x-direction, all values are constant. This means that there is no heat flow in the x-direction. Under the given conditions, after transformation into the frequency domain, Equation 6.2.35 becomes

$$\nabla^2 \underline{\Delta T}_i(y, z, \omega_{Ch}) - \frac{j\omega_{Ch}}{a_i} \underline{\Delta T}_i(y, z, \omega_{Ch}) = 0 \qquad (6.2.36)$$

For Equation 6.2.36, there is a general solution approach using hyperbolic functions:

$$\underline{\Delta T}_i(y, z, \omega_{Ch}) = \hat{E}_S \cos(\omega_y y) e^{j\omega_{Ch}t} [C_{1i} \sinh{(q_i z)} + C_{2i} \cosh{(q_i z)}] \qquad (6.2.37)$$

with

$$q_i^2 = \omega_y + j\frac{\omega_{Ch}}{a_i} \qquad (6.2.38)$$

Constants C_{1i} and C_{2i} have to be determined using the boundary conditions. For the model in Figure 6.2.4, we have to calculate a total of three layers (gas 1 with $i = 1$, sensor element with $i = S$ and gas 2 with $i = 2$). The result is an equation set with three equations and six unknown variables. We want to calculate mean temperature change $\Delta T_d(y, \omega)$ in the sensor layer:

$$\Delta T_d(y, \omega) = \frac{1}{d_S} \int_0^{d_S} \Delta T_S(y, z, \omega) dz \qquad (6.2.39)$$

The following boundary conditions apply:

- At the margins of the sensor element ($z_S = 0$ and $z_S = d_S$), the heat flow has to be constant. The radiant flux is fed in at point $z_S = 0$:

$$-\lambda_S \frac{\partial \underline{\Delta T}_S(z_S = 0)}{\partial z_S} = \hat{E}_S \cos{(\omega_y y)} e^{j\omega_{Ch}t} - G_1' \underline{\Delta T}_S(z_S = 0) - \lambda_1 \frac{\partial \underline{\Delta T}_1(z_1 = d_1)}{\partial z_1} \qquad (6.2.40)$$

$$-\lambda_S \frac{\partial \underline{\Delta T}_S(z_S = d_S)}{\partial z_S} = G_2' \underline{\Delta T}_P(z_S = d_S) - \lambda_2 \frac{\partial \underline{\Delta T}_2(z_2 = 0)}{\partial z_2} \qquad (6.2.41)$$

Here, G_1' and G_2' are the radiation conductances at the detector element surfaces normalised to the sensor area. As the temperature differences on both sensor element areas are small in relation to absolute temperature T_0, the approximation applies:

$$G_1' = G_2' = 4\sigma T_0^3$$

- The temperature changes at the layer boundaries have to be identical:

$$\underline{\Delta T}_1(z_1 = d_1) = \underline{\Delta T}_S(z_S = 0) \qquad (6.2.42)$$

$$\underline{\Delta T}_2(z_2 = 0) = \underline{\Delta T}_S(z_S = d_S) \qquad (6.2.43)$$

- The temperature changes at the bordering heat sinks are zero:

$$\underline{\Delta T}_1(z_1 = 0) = 0 \qquad (6.2.44)$$

$$\underline{\Delta T}_2(z_2 = d_2) = 0 \tag{6.2.45}$$

Using the boundary conditions and calculating in a somewhat complicated way – but without any mathematical tricks and peculiarities – we arrive at the mean temperature change in the sensor element:

$$\underline{\Delta T}_{\mathrm{d}}(y, \omega_{\mathrm{Ch}}) = \frac{\hat{E}_{\mathrm{S}} \cos{(\omega_y y)} e^{j\omega_{\mathrm{Ch}} t}}{q_{\mathrm{S}}^2 \lambda_{\mathrm{S}} d_{\mathrm{S}}} \frac{A_1 \sinh{(q_{\mathrm{S}} d_{\mathrm{S}})} + A_2[\cosh{(q_{\mathrm{S}} d_{\mathrm{S}})} - 1]}{B_1 \sinh{(q_{\mathrm{S}} d_{\mathrm{S}})} + B_2 \cosh{(q_{\mathrm{S}} d_{\mathrm{S}})}} \tag{6.2.46}$$

$$A_1 = \lambda_{\mathrm{S}} q_{\mathrm{S}} \tag{6.2.46a}$$

$$A_2 = G_2' + \lambda_2 q_2 \coth{(q_2 d_2)} \tag{6.2.46b}$$

$$A_3 = G_1' + \lambda_1 q_1 \coth{(q_1 d_1)} \tag{6.2.46c}$$

$$B_1 = \lambda_{\mathrm{S}} q_{\mathrm{S}} + \frac{A_2 A_3}{\lambda_{\mathrm{S}} q_{\mathrm{S}}} \tag{6.2.46d}$$

$$B_2 = A_2 + A_3 \tag{6.2.46e}$$

$$q_{\mathrm{S}}^2 = \omega_y^2 + j\frac{\omega_{\mathrm{Ch}}}{a_{\mathrm{S}}} \tag{6.2.46f}$$

$$q_1^2 = \omega_y^2 + j\frac{\omega_{\mathrm{Ch}}}{a_1} \tag{6.2.46g}$$

$$q_2^2 = \omega_y^2 + j\frac{\omega_{\mathrm{Ch}}}{a_2} \tag{6.2.46h}$$

In order to calculate the mean temperature of a detector element, we finally have to average over rectangular sensor area A_{S}:

$$\underline{\Delta T}_{\mathrm{m}}(\omega_{\mathrm{Ch}}, \omega_y) = \frac{1}{A_{\mathrm{S}}} \int_{A_{\mathrm{S}}} \underline{\Delta T}_d(y, \omega_{\mathrm{Ch}}) \mathrm{d}A \tag{6.2.47}$$

Applying $\hat{E}_{\mathrm{S}} = \dfrac{\hat{\Phi}_{\mathrm{S}}}{A_{\mathrm{S}}}$, it results that

$$\underline{\Delta T}_{\mathrm{m}}(\omega_{\mathrm{Ch}}, \omega_y) = \frac{\hat{\Phi}_{\mathrm{S}} e^{j\omega_{\mathrm{Ch}} t}}{q_{\mathrm{S}}^2 \lambda_{\mathrm{S}} d_{\mathrm{S}} A_{\mathrm{S}}} \mathrm{si}\,(\pi f_y b) \frac{A_1 \sinh{(q_{\mathrm{S}} d_{\mathrm{S}})} + A_2[\cosh{(q_{\mathrm{S}} d_{\mathrm{S}})} - 1]}{B_1 \sinh{(q_{\mathrm{S}} d_{\mathrm{S}})} + B_2 \cosh{(q_{\mathrm{S}} d_{\mathrm{S}})}} \tag{6.2.48}$$

With Equations 6.2.34 and 6.2.1, it applies for the nominator of the left-hand term on the right side of Equation 6.2.48 that

$$q_S^2 \, \lambda_S \, d_S \, A_S = \omega_{Ch} \, C_{th} \left(\frac{a_S \omega_y^2}{\omega_{Ch}} + j \right) \tag{6.2.49}$$

Equation 6.2.48 then becomes

$$\underline{\Delta T}_m \left(\omega_{Ch}, \omega_y \right) = \frac{\hat{\Phi}_S \, e^{j\omega_{Ch}t}}{\omega_{Ch} \, C_{th}} \frac{\mathrm{si}(\pi f_y b)}{\dfrac{a_S \omega_y^2}{\omega_{Ch}} + j} \frac{A_1 \sinh \left(q_S d_S \right) + A_2 [\cosh \left(q_S d_S \right) - 1]}{B_1 \sinh \left(q_S d_S \right) + B_2 \cosh \left(q_S d_S \right)} \tag{6.2.50}$$

This means that the normalised temperature responsivity in the three-layer model is ($\omega_y = 0$)

$$T_R(\omega_y = 0) = \left| \frac{A_1 \sinh \left(q_S d_S \right) + A_2 [\cosh \left(q_S d_S \right) - 1]}{B_1 \sinh \left(q_S d_S \right) + B_2 \cosh \left(q_S d_S \right)} \right| \tag{6.2.51}$$

We arrive at *MTF* normalising Equation 6.2.50 to a zero spatial frequency ($\omega_y = 0$):

$$MTF \, (f_y) = \left| \frac{\underline{\Delta T}_m (f_y)}{\underline{\Delta T}_m (f_y = 0)} \right| = \frac{|\Delta T_m (f_y)|}{T_R(f_y = 0)} \tag{6.2.52}$$

Sensor *MTF* then becomes

$$MTF \, (f_y) = |\mathrm{si}(\pi f_y b)| \frac{1}{\sqrt{1 + \left(\dfrac{a_S \omega_y^2}{\omega_{Ch}} \right)^2}} \frac{\left| \dfrac{A_1 \sinh \left(q_S d_S \right) + A_2 [\cosh \left(q_S d_S \right) - 1]}{B_1 \sinh \left(q_S d_S \right) + B_2 \cosh \left(q_S d_S \right)} \right|}{T_R \, (f_y = 0)} \tag{6.2.53}$$

As Equation 6.2.53 shows, sensor *MTF* consists of three terms. The left-hand term results from integrating over the sensor area according to Equation 6.2.46 and represents geometrical *MTF* (Equation 5.6.22):

$$MTF_g(f_y) = |\mathrm{si} \, (\pi f_y b)| \tag{6.2.54}$$

Both right-hand terms describe the decrease of contrast with increasing spatial frequency due to a heat flow in the detector element. They represent thermal *MTF*:

$$MTF_{th}(f_y) = \frac{1}{\sqrt{1 + \left(\dfrac{a_S \omega_y^2}{\omega_{Ch}} \right)^2}} \frac{\left| \dfrac{A_1 \sinh \left(q_S d_S \right) + A_2 [\cosh \left(q_S d_S \right) - 1]}{B_1 \sinh \left(q_S d_S \right) + B_2 \cosh \left(q_S d_S \right)} \right|}{T_R(f_y = 0)} \tag{6.2.55}$$

This way it applies for the effective value of the mean temperature change that

$$\Delta T_m(\omega_{Ch}, \omega_y) = \frac{\Phi_S}{\omega_{Ch} C_{th}} MTF_g(\omega_y) MTF_{th}(\omega_{Ch}, \omega_y) \tag{6.2.56}$$

Example 6.4: Influence of the Gas Layer on the Normalised Temperature Responsivity

Small device dimensions require miniaturised sensors. Therefore, also the sensor housing has to be as small as possible. This results in very small distances between sensor window or sensor bottom and sensor element. In this example, we want to analyse the influence that the distance of the sensor bottom has on the normalised temperature responsivity. As the sensor element we assume a pyroelectric chip consisting of lithium tantalate $LiTaO_3$. The surrounding gas is assumed to be dry nitrogen. Table 6.2.2 summarizes all necessary parameters.

Figure 6.2.8 presents the normalised temperature responsivity in relation to the chopper frequency.

Using the simple thermal model in Section 6.2.1, the following heat conductance values result (sensor area $A_S = 100 \, \mu m \times 100 \, \mu m = 1 \times 10^{-8} \, m^2$):

- Conductance through radiation: $G_{th,R} = 4 \, \sigma \, T^3 \, A_S = 6.2 \times 10^{-8} \, W/K$
- Conductance through Layer 1: $G_{th,1} = \dfrac{\lambda_{th,Gas}}{d_1} A_S = 2.6 \times 10^{-7} \, W/K$

Table 6.2.2 Parameters used in the example

Layer	Thickness d [µm]	Thermal diffusivity a [m²/s]	Thermal conductivity λ [W/(m K)]
Sensor element	10	1.4×10^{-6}	4.2
Gas 1	1000	2.2×10^{-5}	0.026
Gas 2	300	2.2×10^{-5}	0.026

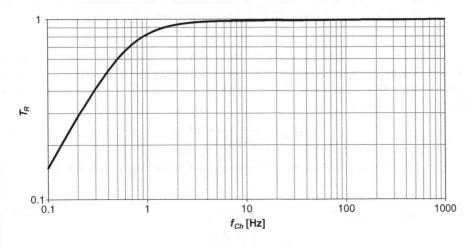

Figure 6.2.8 Normalised temperature responsivity

- Conductance through Layer 2: $G_{th,2} = \dfrac{\lambda_{th,Gas}}{d_2} A_S = 8.6 \times 10^{-7} \text{ W/K}$

- Total conductance: $G_{th} = G_{th,R} + G_{th,1} + G_{th,2} = 1.2 \times 10^{-6} \text{ W/K}$

We arrive at the normalised temperature responsivity:

$$T_R = \frac{G_{th,R}}{G_{th}} = 0.05$$

The heat capacity of the sensor element is $(c_p' = 3.1 \times 10^6 \text{ W s m}^{-3} \text{ K}^{-1})$:

$$C_{th} = c_p' \, d_S \, A_S = 3.1 \times 10^{-7} \text{ Ws/K}$$

The thermal time constant becomes

$$\tau_{th} = \frac{C_{th}}{G_{th}} = 0.28 \text{ s}$$

The layer model is not defined for a modulation frequency f_{Ch} of zero (constant light). The results of the simple thermal model and the thermal layer model are in the same order. We can clearly see the effect of the heat capacity of the sensor element at low frequencies. Thus, it applies for $f_{Ch} = 1$ Hz, for instance:

$$(G_{th} = 1.2 \times 10^{-6} \text{ W/K}) < (\omega_{Ch} C_{th} = 1.9 \times 10^{-6} \text{ W/K}) \tag{6.2.57}$$

If the sensor element is situated in the vacuum, there is no heat conduction through the adjacent layers. With $\lambda_1 = \lambda_2 = 0$, we apply Equation 6.2.50 and arrive at the mean temperature change in the sensor layer:

$$\underline{\Delta T}_m = \frac{\hat{\Phi}_S e^{j\omega_{Ch}t}}{q_S^2 \lambda_S d_S A_S} \text{si}(\pi f_y b) \frac{\lambda_S q_S \sinh(q_S d_S) + G_2' [\cosh(q_S d_S) - 1]}{\left[\lambda_S q_S + \dfrac{G_1' G_2'}{\lambda_S q_S}\right] \sinh(q_S d_S) + \left[G_1' + G_2'\right] \cosh(q_S d_S)} \tag{6.2.58}$$

It becomes obvious that the heat flow in the detector element through the heat capacity often is substantially larger than that of the radiation (see Example 6.4 Table 6.2.2). Thus, it is possible to neglect the conductance through radiation too $(G_1' = G_2' \to 0)$ and we arrive at a frequently used equation:

$$\underline{\Delta T}_m = \frac{\hat{\Phi}_S e^{j\omega_{Ch}t}}{q_S^2 \lambda_S d_S A_S} \text{si}(\pi f_y b) \tag{6.2.59}$$

We can use Equation 6.2.49 to extract amplitude and phase from Equation 6.2.59:

$$\underline{\Delta T}_m = \frac{\hat{\Phi}_S}{\omega_{Ch} C_{th}} \text{si}(\pi f_y b) \frac{1}{\sqrt{1 + \left(\dfrac{a_S \omega_y^2}{\omega_{Ch}}\right)^2}} e^{j\left[\omega_{Ch}t - \arctan\left(\frac{\omega_{Ch}}{a_S \omega_y^2}\right)\right]} \tag{6.2.60}$$

For spatial frequency $f_y = 0$ (constant illumination of the sensor element), the amplitude of the mean temperature change is

$$\Delta T_m \left(\omega_y = 0 \right) = \frac{\Phi_S}{\omega_{Ch} \, C_{th}} \qquad (6.2.61)$$

Comparing Equation 6.2.61 with Equation 6.2.11, the above restriction does become clear. The heat flow through the thermal capacity of the sensor element is substantially larger than the heat flow to the thermal ground:

$$G_{th} \ll \omega_{Ch} \, C_{th} \qquad (6.2.62)$$

For Equation 6.2.60 to be valid, chopper frequency f_{Ch} has to be sufficiently large! The thermal *MTF* of a self-supporting detector element becomes

$$MTF_{th} = \frac{1}{\sqrt{1 + \left(\dfrac{a_S \omega_y^2}{\omega_{Ch}} \right)^2}} \qquad (6.2.63)$$

Example 5.14 presents thermal *MTF* of a self-supporting detector element. It is also interesting to look at the phase of the mean temperature change:

$$\varphi_m = \arctan \left(\frac{\omega_{Ch}}{a_S \omega_y^2} \right) \qquad (6.2.64)$$

It depends on both chopper frequency and spatial frequency. This means that the spatial distribution of the radiant flux affects the point in time of the maximum temperature change in the detector element.

Example 6.5:   Thermal *MTF* for Rectangular Chopping

We mainly use mechanical choppers to modulate the radiation. These so-called multi-bladed wheel choppers generate an almost rectangular radiation-time curve. We can use FOURIER series formation in order to calculate the time signal of the temperature change from the harmonic solutions [30] which contains an elegant solution for the *MTF* of self-supporting detector elements for rectangular modulation. Using FOURIER series formation, we calculate *MTF* for rectangular modulation from Equation 6.2.60:

$$MTF_{th,rect} = \frac{2}{\pi} \frac{\omega_{Ch}}{a_S \omega_y^2} \tanh \left(\frac{\pi}{2} \frac{a_S \, \omega_y^2}{\omega_{Ch}} \right) \qquad (6.2.65)$$

Figure 6.2.9 shows thermal MTF for sinusoidal and rectangular modulation. When comparing, we have to take into account that for rectangular modulation the amplitude of the temperature change at $\omega_y = 0$ is larger by factor $\pi/2$.

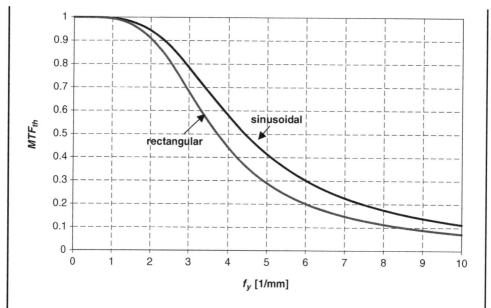

Figure 6.2.9 Thermal *MTF* of a pyroelectric sensor consisting of LiTaO$_3$ with a rectangular and sinusoidal radiation modulation. $f_{\text{Ch}} = 100\,\text{Hz}$; $a_{\text{S}} = 1.4 \times 10^{-6}\,\text{m}^2/\text{s}$

6.3   Network Models for Thermal Sensors

In Section 6.2 we have modelled the thermal relations at thermal IR sensors using thermal conductors and capacities. Analogously to electrical components, these effects can be considered to be thermal components. Here, the conservation of energy (power) and the analysis of temperature differences correspond to KIRCHHOFF's circuit laws in electrical engineering. In the following, we will use these analogies (Table 6.3.1) to derive thermal network models for thermal IR sensors. Here, we are especially interested in representing the thermoelectrical coupling with suitable circuits.

Figure 6.3.1 shows two active two-port networks of thermal sensors. According to the different operating principles, we use either a Z-equivalent (preferably for bolometers and thermoelectric sensors) or a Y-equivalent circuit (preferably for pyroelectric sensors). Both circuit variants can be converted into each other. At the input side, there is a non-electrical, lossy circuit. It is basically the same for all thermal sensors and was calculated in detail in Section 6.2. The electrical output circuit is mainly determined by the integrated signal processing.

Table 6.3.1   Electrothermal analogies

Domain	Flux variables	Potential variables	Components	Power
Electrical	I	φ, V	R, C	$P = VI$
Thermal	Φ	$T, \Delta T$	$R_{\text{th}}, C_{\text{th}}$	$P = \Phi$

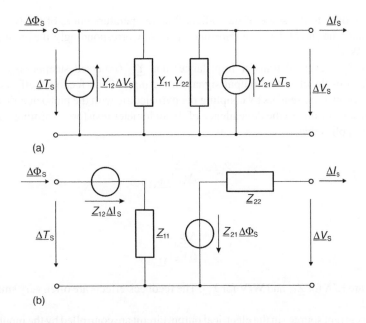

Figure 6.3.1   Description of a thermal sensor with active two-port networks. (a) Y-equivalent two-port. (b) Z-equivalent two-port

In general, the two-port equations are as follows:

- Z-equivalent circuit (resistance or impedance form):

$$\underline{\Delta T}_S = \underline{Z}_{11}\,\underline{\Delta \Phi}_S + \underline{Z}_{12}\,\underline{\Delta I}_S \tag{6.3.1}$$

$$\underline{\Delta V}_S = \underline{Z}_{21}\underline{\Delta \Phi}_S + \underline{Z}_{22}\underline{\Delta I}_S \tag{6.3.2}$$

- Y-equivalent circuit (conductance or admittance form):

$$\underline{\Delta \Phi}_S = \underline{Y}_{11}\underline{\Delta T}_S + \underline{Y}_{12}\underline{\Delta V}_S \tag{6.3.3}$$

$$\underline{\Delta I}_S = \underline{Y}_{12}\underline{\Delta T}_S + \underline{Y}_{22}\underline{\Delta V}_S \tag{6.3.4}$$

All two-port parameters are complex variables that can be easily calculated using the two-port equations. It applies for input impedance $\underline{Z}_{11}$ or input admittance $\underline{Y}_{11}$ that

$$\underline{Z}_{11} = \left.\frac{\underline{\Delta T}_S}{\underline{\Delta \Phi}_S}\right|_{\underline{\Delta I}_S=0} \tag{6.3.5}$$

or

$$\underline{Y}_{11} = \left.\frac{\underline{\Delta \Phi}_S}{\underline{\Delta T}_S}\right|_{\underline{\Delta V}_S=0} \tag{6.3.6}$$

This corresponds to the slope of the radiant flux–temperature curve of the two-port at the operating point. The unit of the input impedance is K/W. Correspondingly, the input admittance has the unit W/K.

Parameter $\underline{Z}_{12}$ (reversed transfer impedance) and $\underline{Y}_{12}$ (reversed steepness) describe the feedback effect of the electrical output signal on the input. Such feedback effects occur, for instance, in pyroelectric sensors by coupling the pyroelectric with the piezoelectric effect or in bolometrical sensors due to the dependence of the bolometer resistor's warming on operating current I_0. It applies that

$$\underline{Z}_{12} = \frac{\Delta T_S}{\Delta I_S}\bigg|_{\Delta\Phi_S=0} \tag{6.3.7}$$

or

$$\underline{Y}_{12} = \frac{\Delta\Phi_S}{\Delta V_S}\bigg|_{\Delta T_S=0} \tag{6.3.8}$$

The units are K/A for $\underline{Z}_{12}$ and W/V for $\underline{Y}_{12}$. The feedback effects are often very small and can be neglected.

Voltage or current sources in the electrical output circuit are controlled by the input variables. Responsivity or steepness can be described by two-port parameters $\underline{Z}_{21}$ and $\underline{Y}_{21}$:

$$\underline{Z}_{21} = \frac{\Delta V_S}{\Delta\Phi_S}\bigg|_{\Delta I_S=0} \tag{6.3.9}$$

or

$$\underline{Y}_{21} = \frac{\Delta I_S}{\Delta T_S}\bigg|_{\Delta V_S=0} \tag{6.3.10}$$

Two-port parameter $\underline{Z}_{21}$ corresponds to voltage responsivity R_V with the unit V/W. Parameter $\underline{Y}_{21}$ has the unit A/K.

Output impedance $\underline{Z}_{22}$ and output admittance $\underline{Y}_{22}$ are calculated as follows:

$$\underline{Z}_{22} = \frac{\Delta V_S}{\Delta I_S}\bigg|_{\Delta\Phi_S=0} \tag{6.3.11}$$

or

$$\underline{Y}_{22} = \frac{\Delta I_S}{\Delta U_S}\bigg|_{\Delta T_S=0} \tag{6.3.12}$$

The unit of output impedance $\underline{Z}_{22}$ is Ω and output admittance $\underline{Y}_{22}$ has the unit S.

We can also use passive two-ports for describing thermal IR sensors that operate according to the principle of energy conversion (pyroelectric and thermoelectric sensors). This method is useful if we want to analyse the internal relations within the sensor. An example here is the coupling of the pyroelectric and piezoelectric effect. Due to the reversibility of the two-ports, it applies in this case $\underline{Z}_{12}=\underline{Z}_{21}$ or $\underline{Y}_{12}=\underline{Y}_{21}$. However, it is better to use active two-ports to describe sensor behaviour in interaction with its environment. As, in principle, reversibility does not apply to parametric transducers, in the following we will consistently use equivalent circuits with active sources.

Example 6.6: Two-Port Parameters of a Microbolometer

Figure 6.3.2 shows the small-signal circuit of a microbolometer. In the non-electrical input circuit, the microbolometer is represented by its heat capacity C_{th} and its heat resistance R_{th}. At the output side, there is open-circuit voltage $\Delta \underline{V}_S$ modulated by input flux $\Delta \Phi_S$. The bolometer resistance is R_B. We will start by calculating the impedance matrix.

For bolometers as parametric transducers, there is usually no feedback of sensor current $\Delta \underline{I}_S$ on temperature difference $\Delta \underline{T}_S$, therefore it has to apply that

$$\underline{Z}_{12} = 0 \tag{6.3.13}$$

For this reason, there is no controlled temperature source in the left-hand input circuit. Input impedance $\underline{Z}_{11}$ is

$$\underline{Z}_{11} = R_{th} \left\| \frac{1}{j\omega C_{th}} = \frac{R_{th}}{1+j\omega C R_{th}} \right. \tag{6.3.14}$$

and thus represents the thermal properties (heat conduction and heat capacity) of the sensor element.

For the open circuit ($\Delta \underline{I}_S = 0$), parameter $\underline{Z}_{21}$ is calculated as the quotient of open-circuit output voltage $\Delta \underline{V}_S$ and the absorbed radiant flux difference $\Delta \Phi_S$ and corresponds to the voltage responsivity:

$$\underline{Z}_{21} = R_V \tag{6.3.15}$$

The output impedance is determined through bolometer resistance R_B:

$$\underline{Z}_{22} = R_B \tag{6.3.16}$$

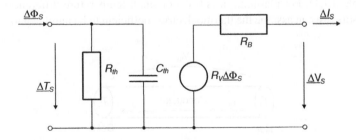

Figure 6.3.2 Small-signal equivalent circuit of a microbolometer

With the calculated parameters, the impedance matrix becomes

$$\underline{Z} = \begin{pmatrix} \dfrac{R_{th}}{1+j\omega C_{th}R_{th}} & 0 \\ R_V & R_B \end{pmatrix} \tag{6.3.17}$$

According to the calculation rule

$$\underline{Y} = \underline{Z}^{-1} = \frac{1}{\det \underline{Z}} \begin{vmatrix} \underline{Z}_{22} & -\underline{Z}_{12} \\ -\underline{Z}_{21} & \underline{Z}_{11} \end{vmatrix} \tag{6.3.18}$$

we can use it to calculate the admittance matrix:

$$\underline{Y} = \begin{vmatrix} G_{th}+j\omega C_{th} & 0 \\ -R_V \dfrac{1+j\omega C_{th}R_{th}}{R_{th}R_B} & \dfrac{1}{R_B} \end{vmatrix} \tag{6.3.19}$$

6.4   Thermoelectric Radiation Sensors

6.4.1   *Principle*

Thermoelectric sensors utilise thermoelectric voltage V_{th} occurring in a conductor for measuring temperature difference $\Delta T_S = T_1 - T_2$ between sensor element (measuring point with temperature T_1, hot junction) and a reference temperature (reference point with temperature T_2, cold junction) (Figure 6.4.1).

The thermoelectric voltage is caused by thermal diffusion of charge carriers (thermal diffusion current). The thermoelectric effect is also called SEEBECK effect. Thermoelectric voltage V_{th} is directly proportional to temperature difference ΔT_S:

$$V_{th} = \alpha_S(T_1 - T_2) = \alpha_S \Delta T_S \tag{6.4.1}$$

It only depends on temperature difference ΔT_S and not on the spatial curve of the temperature! The proportionality factor is thermoelectric coefficient α_S, also called SEEBECK coefficient or thermal power. It has the unit V/K. Thermoelectric coefficient α_S is a material constant (Table 6.4.1). For platinum, it is 0 V. For small temperature differences, such as in radiation sensors, we can assume the thermoelectric coefficient to be unrelated to temperature.

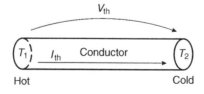

Figure 6.4.1   Thermoelectric effect

Table 6.4.1 Thermoelectric voltage series at a reference temperature of $0\,^{\circ}C$ and a temperature difference of $100\,K$

Substance	V_{th} [mV]	Substance	V_{th} [mV]
Tellurium	$+50$	Platinum	0
Nickel-chromium	$+2.2$	Sodium	-0.21
Iron	$+1.88$	Nickel	$-1.94 \ldots 1.2$
Cadmium	$+0.85 \ldots 0.92$	Constantan	$-3.47 \ldots 3.04$
Copper	$+0.72 \ldots 0.77$	Bismuth	-7
Gold	$+0.56 \ldots 0.8$		
Aluminium	$+0.37 \ldots 0.41$		

The PELTIER effect and the THOMSON effect are closely related to the SEEBECK effect. The PELTIER effect is the reversal of the SEEBECK effect. The THOMSON effect describes the heat conduction caused by a flowing current. The interrelations between the transport of thermal energy and electrical energy in an electrical conductor can be described mathematically as follows [1]:

$$\begin{pmatrix} J \\ J_{th} \end{pmatrix} = \begin{pmatrix} \kappa T & \kappa \alpha_S T^2 \\ \kappa \alpha_S T^2 & \lambda T^2 + \kappa \alpha_S^2 T^3 \end{pmatrix} \begin{pmatrix} \dfrac{E}{T} \\ -\dfrac{\nabla T}{T^2} \end{pmatrix} \tag{6.4.2}$$

with J being the electric current density, J_{th} the heat flux density, κ the electrical and λ the thermal conductivity. For a constant temperature, OHM'S law results from Equation 6.4.2:

$$J = \kappa E \tag{6.4.3}$$

or with $I = J A$ and $V = E l$ in its familiar form:

$$R = \frac{V}{I} = \frac{l}{\kappa A} \tag{6.4.4}$$

For a currentless condition $(J = 0)$ and with temperature gradient ∇T, Equation 6.4.2 becomes FOURIER'S heat transfer law:

$$J_{th} = -\lambda \, \nabla T \tag{6.4.5}$$

An electric current also causes a heat flux (PELTIER effect). For $\nabla T = 0$, it results from Equation 6.4.2:

$$J_{th} = \alpha_S T \, J \tag{6.4.6}$$

The reason for the thermoelectric effect lies in the statistics of free charge carriers. For non-degenerate semiconductors, we can calculate the SEEBECK coefficients as well as the conductivity using MAXWELL–BOLTZMANN statistics. Here, the SEEBECK effect states the fact that a temperature gradient in a (semi-)conductor results in a gradient of the FERMI level:

$$\frac{\nabla E_F}{q} = \alpha_S \nabla T \tag{6.4.7}$$

with the energy of FERMI level E_F (FERMI energy) and charge q. The SEEBECK coefficient thus becomes

$$\alpha_S = \frac{1}{q}\frac{dE_F}{dT} \tag{6.4.8}$$

This confirms that the SEEBECK coefficient does not depend on the spatial distribution of the temperature gradient, but only on the temperature difference. The FERMI level for n-semiconductors is

$$E_F = E_C - k_B T \ln\frac{N_C}{n} \tag{6.4.9}$$

and for p-semiconductors

$$E_F = E_V + k_B T \ln\frac{N_V}{n} \tag{6.4.10}$$

with density of states N_C and N_V in the conduction band and in the valence band, respectively, number of charge carriers n in the conduction band and band edges E_C and E_V of the conduction band and the valence band, respectively. We have to take into account for the calculation of the SEEBECK coefficient that the intrinsic conduction increases with increasing temperature, that the mobility of the charge carriers increases and that there occurs a phonon flux from warm to cold. It thus applies for non-degenerated semiconductors according to [8] with $q=e$

$$\alpha_{S,n} = -\frac{k_B}{e}\left[\ln\frac{N_C}{n} + \frac{3}{2} + (1+s_n) + \Phi_n\right] \tag{6.4.11}$$

and for p-semiconductions with $q=-e$

$$\alpha_{S,p} = \frac{k_B}{e}\left[\ln\frac{N_V}{p} + \frac{3}{2} + (1+s_p) + \Phi_p\right] \tag{6.4.12}$$

Here, the two first terms

$$k_B\left(\ln\frac{N_C}{n} + \frac{3}{2}\right) = S_n$$

and

$$k_B\left(\ln\frac{N_V}{p} + \frac{3}{2}\right) = S_p$$

are correspondingly the entropy per electron or per hole with densities of states N_C and N_L and the number of charge carriers n or p.

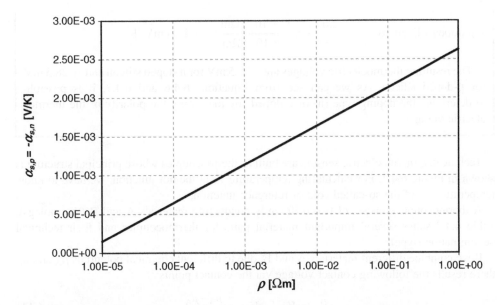

Figure 6.4.2 SEEBECK coefficients $\alpha_{s,p}$ and $\alpha_{s,n}$, respectively, of doped single-crystal silicon versus electrical conductivity ρ

$(1 + s_n)$ or $(1 + s_p)$ take into consideration the different mobilities of the charge carriers between warm and cold conductor end $(s_{n,p} = -1 \ldots +2)$. Thus, s_n or s_p describe the ratio of relaxation time and charge carrier energy. Φ_n or Φ_p describe the phonon drags that move from the warm to the cold conductor end $(\Phi_{n,p} = 0 \ldots 5)$.

For real-life applications, Equations 6.4.11 and 6.4.12 can be approximated using a simple formula [3]:

$$\alpha_S = \frac{m \, k_B}{q} \ln \frac{\rho}{\rho_0} \tag{6.4.13}$$

with constants $m \approx 2.5$ and $\rho_0 \approx 5 \times 10^{-6} \, \Omega \, m$ for single-crystal silicon and electrical conductivity ρ. With $q = e$, it results that $m k_B / q \approx 0.22 \, \text{mV/K}$. Figure 6.4.2 presents this correlation.

Example 6.7:   Thermoelectric Voltage at Doped Silicon

We locally heat an n- and a p-conducting silicon wafer and measure the voltage between the heated and the non-heated part. The temperature difference is 25 K, conductivity of silicon is 1 mΩ m. With Equation 6.4.8, the following SEEBECK coefficients result:

- n-doped silicon: $\alpha_{S,n} = -\dfrac{2.6 \, k_B}{e} \ln \dfrac{1 \times 10^{-3} \, \Omega m}{5 \times 10^{-6} \, \Omega m} = -1.14 \, \text{mV/K}$

- p-doped silicon: $\alpha_{S,p} = -\dfrac{2.6\,k_B}{-e}\ln\dfrac{1\times10^{-3}\,\Omega\text{m}}{5\times10^{-6}\,\Omega\text{m}} = +1.14\,\text{mV/K}$

The resulting thermoelectric voltages are $-28.5\,\text{mV}$ for n-doped silicon and $+28.5\,\text{mV}$ for p-doped silicon. As we can see from Equations 6.4.6 and 6.4.7, it is possible to determine the doping type (n- or p-doped) by measuring the polarity of the thermo-electric voltage.

Technically, thermoelectric sensors are built as thermocouples whose principal structure is shown in Figure 6.4.3. For calculating temperature T_1 at the hot junction, we have to know temperature T_2 of the so-called cold or reference junction.

A thermocouple consists of two differently conducting materials A and B (thermolegs). Table 6.4.2 states several important material pairs for thermocouples and their technical denomination (type).

If we combine different materials A and B, due to different chemical potentials μ_A and μ_B there results the following contact voltage via the contact points:

$$V_K = V_{K1} = V_{K2} = \frac{\mu_B - \mu_A}{q} \tag{6.4.14}$$

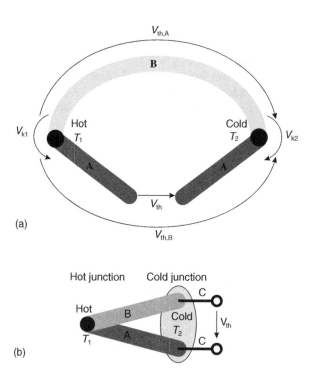

(a)

(b)

Figure 6.4.3 Principle of thermocouples: (a) general arrangement; (b) common technical implementation (material $C = A$ or B)

Table 6.4.2 SEEBECK coefficients of technically important thermocouples

Material pairs A/B acc. Figure 6.4.3	Type	$\alpha_{S,AB}$ at 300 K [μV/K]
Fe/Cu-Ni	J	51
Ni-Cr/Ni-Al	K	41
Pt-13%Rh/Pt	R	6
Pt-10%Rh/Pt	S	7
Cu/Cu-Ni	T	41

with charge q (in metals due to electron conduction $q = e$). In the relevant temperature range, we can usually assume the contact voltage to be unrelated to temperature. For the thermo-electric voltage of a thermocouple consisting of two electrical conductors, it applies according to Equation 6.4.1 that

$$V_{th,A} = \alpha_{S,A}(T_1 - T_2) \tag{6.4.15}$$

or

$$V_{th,B} = \alpha_{S,B}(T_1 - T_2) \tag{6.4.16}$$

and thus

$$V_{th} = V_{K2} + V_{th,A} - V_{K1} - V_{th,B} \tag{6.4.17}$$

As the two contact voltages are identical, Equation 6.4.17 becomes

$$V_{th} = V_{th,A} - V_{th,B} = (\alpha_{S,A} - \alpha_{S,B})\Delta T_S = \alpha_{S,AB}\Delta T_S \tag{6.4.18}$$

with

$$\alpha_{S,AB} = \alpha_{S,A} - \alpha_{S,B} \tag{6.4.19}$$

This equation shows that we need to use thermocouples with the largest absolute value of SEEBECK coefficients, but with unlike signs. The difference of the thermoelectric coefficients $\alpha_{S,A} - \alpha_{S,B}$ is very small (Tables 6.4.1–6.4.3). Therefore we usually use for infrared sensors many thermocouples in series. Such an arrangement is called thermopile. For N series thermocouples, the measured thermoelectric voltage thus becomes

$$V_{th} = N\alpha_{AB}\,\Delta T_S \tag{6.4.20}$$

If we short-circuit the ends of a thermocouple, we get a thermoelectric current I_{th} that is only limited by resistance R_{th} of the thermocouple and can become very large:

$$I_{th} = \frac{V_{th}}{R_{th}} \tag{6.4.21}$$

Table 6.4.3 Characteristics of thermoelectric materials [4, 5]

	Material	α_S [μV/K]	κ [$10^4\,\Omega^{-1}\,m^{-1}$]	λ [W/(Km)]	FOM [K^{-1}]	M(300 K)
Semi conductors	Si (bulk)	-450	2.85	145	4×10^{-5}	0.012
	n-Si	$-100 \ldots -800$	$0.2 \ldots 10$	150	$6.3 \times 10^{-6} \ldots 9.1 \times 10^{-6}$	$0.0019 \ldots 0.0027$
	n-poly-Si	$-100 \ldots -500$	$0.1 \ldots 10$	$20 \ldots 30$	$1.4 \times 10^{-5} \ldots 5 \times 10^{-5}$	$0.0042 \ldots 0.015$
	p-Si	$100 \ldots 800$	$0.1 \ldots 10$	150	$6.3 \times 10^{-6} \ldots 9.1 \times 10^{-6}$	$1.9 \times 10^{-3} \ldots 2.7 \times 10^{-3}$
	p-poly-Si	$100 \ldots 500$	$0.1 \ldots 10$	$20 \ldots 30$	$1.4 \times 10^{-5} \ldots 5 \times 10^{-5}$	$0.0019 \ldots 0.0027$
	Ge	420	0.12	64	3.3×10^{-6}	0.00099
	Sb	35	100	13	9×10^{-5}	0.027
	Bi	-65	28.5	4.2	2.3×10^{-4}	0.069
	p-$Bi_{0.5}Sb_{1.5}Te_3$	230	5.8	1.05	2.9×10^{-3}	0.87
	n-$Bi_{0.87}Sb_{0.13}$	-100	14	13	4.5×10^{-4}	0.135
	PbTe	-170	5.0	2.5	5.8×10^{-4}	0.174
	$Bi_{1.8}Sb_{0.2}Te_{2.7}Se_{0.3}$	-220	9.1	1.4	3.15×10^{-3}	0.945
Metals	Al	-3.2	3541.4	235	1.54×10^{-6}	4.62×10^{-4}
	Au	0.1	4347.8	315	1.38×10^{-9}	4.14×10^{-7}
	Ni	-20.2	1628.6	61	1.08×10^{-5}	0.00324
	Pt	-3.2	1019.3	71	1.47×10^{-6}	4.4×10^{-4}

6.4.2   Thermal Resolution

The responsivity of a thermocouple results from Equations 6.2.13 and 6.4.18:

$$R_V(f_{Ch}) = \frac{\alpha\,\alpha_{S,AB}}{G_{th}}\,\frac{1}{\sqrt{1+\omega_{Ch}^2\tau_{th}}} \qquad (6.4.22)$$

As thermocouples are constant-light sensitive, we usually state the voltage responsivity $R_V(f_{Ch}=0)$:

$$R_V = \frac{\alpha\,\alpha_{S,AB}}{G_{th}} \qquad (6.4.23)$$

As for all thermal sensors, the dynamic processes are determined by thermal time constant τ_{th}. Neglecting possible substrate layers to which the thermocouples are attached, the thermal conductance can be calculated as

$$G_{th} = G_{th,R} + G_{th,AB} \qquad (6.4.24)$$

with the thermal conductance through radiation $G_{th,S}$ according to Equation 6.1.14 and due to thermal conduction through the legs of the thermocouples:

$$G_{th,AB} = G_{th,A} + G_{th,B} \qquad (6.4.25)$$

The thermal conductance through legs A and B becomes

$$G_{th,A} = \frac{\lambda_A A_A}{l_A} \qquad (6.4.26)$$

$$G_{th,B} = \frac{\lambda_B A_B}{l_B} \qquad (6.4.27)$$

with thermal conductivities λ_A and λ_B of the leg materials, cross-sectional areas A_A and A_B of the legs and leg lengths l_A and l_B.

The most important noise source of the thermocouples is the thermal resistance noise of the thermocouple

$$\tilde{v}_{Rn}^2 = 4k_B T\,R_{AB} \qquad (6.4.28)$$

with resistance R_{AB}

$$R_{AB} = \frac{l_A}{A_A\kappa_A} + \frac{l_B}{A_B\kappa_B} \qquad (6.4.29)$$

κ_A and κ_B are the electrical conductivities of the leg materials. Specific *NEP* thus becomes

$$NEP_{TE}^* = \frac{\sqrt{4k_B\,T\,R_{AB}}}{\alpha\,\alpha_{S,AB}}\,G_{th} \qquad (6.4.30)$$

NEP contains both thermal and electrical material constants. In order to optimise the sensor design, we state a specific quality factor (Figure of Merit), FOM_{AB}.[3] It contains all important sensor parameters, such as thermoelectric power, geometry, thermal and electrical conductivity:

$$FOM_{AB} = \frac{\alpha_{S,AB}^2}{R_{AB}\,G_{th,AB}} \qquad (6.4.31)$$

Under the condition that the thermal conductivity is determined by that of the thermolegs, it applies for specific *NEP*:

$$NEP^* = \frac{1}{\alpha}\sqrt{\frac{4k_B T\,G_{th,AB}}{FOM_{AB}}} \qquad (6.4.32)$$

Quality factor FOM_{AB} should reach its maximum in order for specific *NEP* to be at its minimum. For maximising FOM_{AB}, we assume that the length of both thermolegs is given by the structure and that the length of both legs is identical:

$$l = l_A = l_B \qquad (6.4.33)$$

Thus, it results from Equation 6.4.31:

$$FOM_{AB} = \frac{\alpha_{AB}^2}{\left(\dfrac{1}{\kappa_A A_A} + \dfrac{1}{\kappa_B A_B}\right)(\lambda_A A_A + \lambda_B A_B)} \qquad (6.4.34)$$

The optimising variable is now the ratio of the leg cross-sections A_A/A_B. In order to maximise it, we calculate the derivative of Equation 6.4.34 with respect to A_A/A_B. The derivative $dFOM_{AB}/d\,(A_A/A_B)$ is set to zero; and after transforming the equation we arrive at

$$\frac{A_A}{A_B} = \sqrt{\frac{\kappa_B \lambda_B}{\kappa_A \lambda_A}} \qquad (6.4.35)$$

We now can use Equation 6.4.34 to calculate the optimum thicknesses and widths of the legs. Inserting Equation 6.4.35 into Equation 6.4.31, we arrive at the geometrical maximum of quality factor $FOM_{AB,max}$:

$$FOM_{AB,max} = \frac{\alpha_{AB}^2}{\left(\sqrt{\dfrac{\lambda_A}{\kappa_A}} + \sqrt{\dfrac{\lambda_B}{\kappa_B}}\right)^2} \qquad (6.4.36)$$

[3] For quality factor FOM_{AB}, we also use the abbreviations z_{AB} or Z_{AB}.

In order to assess materials, we can use quality factor *FOM* for a single material instead of a thermocouple (Table 6.4.3):

$$FOM_{A,max} = \frac{\frac{\alpha_A^2}{\lambda_A}}{\kappa_A} = \frac{\kappa_A\,\alpha_A^2}{\lambda_A} \tag{6.4.37}$$

We now can use Equation 5.3.2 to calculate the specific detectivity based on specific *NEP*:

$$D_{TE}^* = \frac{\alpha\,\alpha_{AB}}{G_{th}}\sqrt{\frac{A_S}{4\,k_B\,T_S\,R_{AB}}} \tag{6.4.38}$$

It applies for the thermal conductance that

$$G_{th} = G_{th,S} + G_{th,AB} = 4\,\alpha\,\sigma T_S^3\,A_S + G_{th,AB} \tag{6.4.39}$$

If we in addition insert quality factor FOM_{AB} according to Equation 6.4.30, we arrive at the specific detectivity:

$$D_{TE}^* = \frac{1}{4}\sqrt{\frac{FOM_{AB}}{k_B\,\sigma\,T_S^4}\,\frac{1}{\left(\dfrac{G_{th,S}}{G_{th,AB}} + \dfrac{G_{th,AB}}{G_{th,S}} + 2\right)}} \tag{6.4.40}$$

For the nominator it was taken into account that the quality factor only uses the thermal conductance $G_{th,AB}$ of the thermocouple legs. The right-hand fraction now has its maximum at

$$G_{th,S} = G_{th,AB} \tag{6.4.41}$$

Considering this design specification, the maximum of the specific detectivity becomes [6]

$$D_{TE,max}^* = \frac{1}{8}\sqrt{\frac{FOM_{AB}}{k_B\,\sigma\,T_S^4}} = \frac{1}{8}\sqrt{\frac{M}{k_B\,\sigma\,T_S^5}} \tag{6.4.42}$$

with

$$M = FOM_{AB}\,T_S \tag{6.4.43}$$

Quality factor M is thus proportional to quality factor FOM_{AB}, which means that both are equivalent. However, M includes sensor temperature T_S. Quality factor M is commonly used as it is a dimensionless number and therefore suitable as an easy-to-use comparison parameter. If we now compare Equation 6.4.42 with the BLIP detectivity of thermal sensors according to Equation 5.3.4, it results for value M

$$M \le 4 \tag{6.4.44}$$

It thus applies that

$$D_{TE,max}^* = \frac{\sqrt{M}}{2}\,D_{BLIP}^* \tag{6.4.45}$$

As already mentioned, we combine several thermocouples into thermopiles in order to increase responsivity. In addition to the thermoelectric voltage that increases by factor N according to Equation 6.4.20, the resistance and consequently also the noise increase by factor $\sqrt{N}$. Even the thermal conductance may increase. However, assuming that it does not increase and applying Equation 6.4.30, we arrive at specific NEP^*_{TP} for a thermopile:

$$NEP^*_{TP} = \frac{\sqrt{4\,k_B\,T\,N\,R_{AB}}}{\alpha\,N\,\alpha_{AB}} = \frac{1}{\sqrt{N}} NEP^*_{TE} \qquad (6.4.46)$$

With Equation 6.4.38, it applies for the specific detectivity of a thermopile that

$$D^*_{TP} = \sqrt{N}\,D^*_{TE} \qquad (6.4.47)$$

This means that the signal-to-noise ratio improves with the square root of number N of the series thermocouples.

Equations 5.4.6, 5.4.8 and 6.4.47 can be used to calculate temperature resolution $NETD_{TP}$ from the specific detectivity:

$$NETD_{TP} = 8\frac{4\,k^2 + 1}{I_M}\sqrt{\frac{k_B\,\sigma\,T_S^4\,B}{A_S N\,z_{AB}}} \qquad (6.4.48)$$

6.4.3   Design of Thermoelectric Sensors

In order to achieve a sufficiently large output voltage, in practice we often design thermoelectric radiation sensors as thermopiles (Figure 6.4.4). For thermal isolation, the hot junctions of the thermocouple are applied to a thin membrane. The cold junctions are situated on the silicon

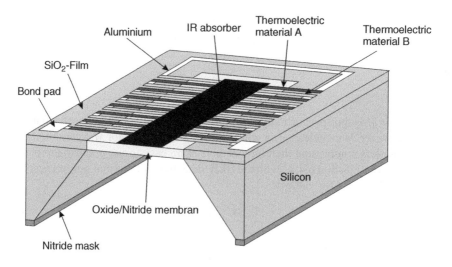

Figure 6.4.4   Principal structure of a thermopile

carrier that – because of its favourable conductivity – constitutes a thermal mass with a constant temperature. The thermopiles are manufactured using standard microelectronics technology. At first, the electronics required at the front side of the Si-wafer is completed, usually in a CMOS-compatible process. After that the thermocouples are carried out. Finally, the membranes are wet-chemically etched from the backside [3].

Example 6.8: Dimensioning of Thermopiles

We have to determine the temperature resolution for a thermopile according to Figure 6.4.4. For this purpose, first we have to dimension the thermocouple. The following conditions must apply:

- Sensor area A_S is 1.0 mm × 0.5 mm. It is determined by the size of the absorber area.
- As thermocouple, we select n-poly-Si/p-poly-Si. There will be 16 thermocouples, eight on each side. The following material parameters are given:
    - SEEBECK coefficient $\alpha_{p\text{-Poly}} = -\alpha_{n\text{-Poly}} = 300\,\mu\text{V/K}$; $\alpha_{AB} = 600\,\mu\text{V/K}$
    - Thermal conductivity $\lambda_{p\text{-Poly}} = \lambda_{n\text{-Poly}} = 25\,\text{W/(Km)}$
    - Electrical conductivity $\kappa_{p\text{-Poly}} = \kappa_{n\text{-Poly}} = 5 \times 10^4\,\Omega^{-1}\,\text{m}^{-1}$
- Thickness d_M of the silicon oxide/nitride membrane is 1 μm, its area $A_M = 4\,\text{mm} \times 2\,\text{mm}$. Thermal conductivity λ_M of the membrane is assumed to be approximately 1 W/(Km).

Given these assumptions, we can now calculate quality factors FOM_{AB} and M according to Equations 6.4.36 and 6.4.43:

- $FOM_{AB} = 1.8 \times 10^{-4}\,\text{K}^{-1}$
- $M(300\,\text{K}) = 0.054$

We have to install eight thermocouples on each side. For an edge length of the sensor area of 1 mm, each thermocouple can take up a width of 0.125 mm. The minimum width and thickness of the legs is limited by technology. Here we will choose width of $b = 10\,\mu\text{m}$. Both legs are assumed to have an identical width. Length l of the legs to the heat sink on the membrane is approximately 1 mm. It follows from Equation 6.4.35 that the leg cross-sections should be equal.

With the given sensor area, we can use Equation 6.1.14 to calculate the temperature conductance through radiation:

$$G_{th,S} = 4\,\sigma_S(300\,\text{K})^3 0.5\,\text{mm}^2 = 3.1 \times 10^{-6}\,\text{W/K} \qquad (6.4.49)$$

The thermal conductance of the membrane can be estimated for a thermocouple as follows:

$$G_{th,TC} = \frac{\lambda_M A_M}{l_M} \qquad (6.4.50)$$

Area A_M results from the membrane's thickness of 1 μm and from the width that each thermocouple can take up (in this example 0.125 mm). Length l_M is assumed to correspond to the distance between sensor area and Si-carrier ($l_M = 2.5\,\text{mm}$). Thus it applies that

$$G_{\text{th,TC}} = \frac{1\,\dfrac{\text{W}}{\text{Km}} \cdot 1.25 \times 10^{-10}\,\text{m}^2}{2.5\,\text{mm}} = 5 \times 10^{-8}\,\text{W/K} \qquad (6.4.51)$$

The total conductance is then

$$G_{\text{th,M}} = 16\,G_{\text{th,TC}} = 8 \times 10^{-7}\,\text{W/K} \qquad (6.4.52)$$

That means that it is considerably smaller than the thermal conductance through radiation and can therefore be neglected.

Using the conditions in Equation 6.4.41, we can calculate the thickness of the thermo-legs. Each thermocouple contributes the same amount to the heat conduction. The number of legs is 32. That means it applies for each leg that:

$$d = \frac{G_{\text{th,S}}}{32}\frac{l}{b\,\lambda_{\text{Poly-Si}}} = \frac{3.1 \times 10^{-6}\,\text{W/K}}{32}\frac{1\,\text{mm}}{10\,\mu\text{m} \times 25\text{W}/(\text{Km})} = 0.39\,\mu\text{m} \qquad (6.4.53)$$

It thus applies for total thermal conductance G_{th} that

$$G_{\text{th}} = 2\,G_{\text{th,S}} = 6.2 \times 10^{-6}\,\text{W/K} \qquad (6.4.54)$$

From Equation 6.4.42, it results for the specific detectivity that

$$D^*_{\text{TE}} = \frac{1}{8}\sqrt{16\frac{1.8 \times 10^{-4}\,\text{K}^{-1}}{k_\text{B}\,\sigma_\text{B}(300\text{K})^4}} = 8.4 \times 10^7\,\frac{\text{m}\sqrt{\text{Hz}}}{\text{W}} \qquad (6.4.55)$$

The temperature resolution results from f-number $F = 1$, specific bandwidth $B = 100\,\text{Hz}$ that is determined by the electronics, and a wavelength range of 8–14 μm (value of differential exitance $I_M = 2.64\,\text{W}/(\text{Km}^2)$) according to Equation 6.4.48:

$$NETD = 8\frac{5}{2.64\,\dfrac{\text{W}}{\text{m}^2\text{K}}}\sqrt{\frac{k_\text{B}\,\sigma_\text{B}(300\text{K})^4 \times 100\,\text{Hz}}{0.5\,\text{mm}^2 \times 16 \times 1.8 \times 10^{-4}\,\text{K}^{-1}}} = 0.3\,\text{mK} \qquad (6.4.56)$$

Finally, we estimate thermal time constant τ_{th}. For this, we still have to calculate heat capacity C_{th}. It is the sum of the heat capacities of the 32 thermocouple legs and the membrane:

$$C_{\text{th}} = \sum_{i=1}^{32} c_i\,V_i + c_\text{M}V_\text{M} \qquad (6.4.57)$$

with volume-specific heat capacity c_i or c_M and volume V_i or V_M of layer i or M of the membrane. The volume-specific heat capacity of poly-Si is $1.6 \times 10^6\,\text{Ws}/(\text{m}^3\text{K})$ and for the silicon oxide/nitride membrane approximately 2×10^6 Ws/(m^{3}K). The volumes amount to

$$V_i = d\,b\,l = 0.39\,\mu\text{m} \times 10\,\mu\text{m} \times 1\,\text{mm} = 3.9 \times 10^{-15}\,\text{m}^3 \qquad (6.4.58)$$

and

$$V_M = 1\,\text{mm} \times 0.5\,\text{mm} \times 1\,\mu\text{m} = 5 \times 10^{-13}\,\text{m}^3 \qquad (6.4.59)$$

Thus, the heat capacity becomes

$$C_{th} = \sum_{i=1}^{32} 6.24 \times 10^{-9}\,\text{Ws/K} + 1 \times 10^6\,\text{Ws/K} = 1.2 \times 10^{-6}\,\text{Ws/K} \qquad (6.4.60)$$

It becomes clear that the heat capacity of the membrane cannot be neglected due to its volume. The resulting thermal time constant is

$$\tau_{th} = \frac{C_{th}}{G_{th}} = \frac{1.2 \times 10^{-6}\,\text{Ws/K}}{36.2 \times 10^{-6}\,\text{W/K}} = 0.195\,\text{s} \qquad (6.4.61)$$

In order to evaluate the thermoelectric voltage, we need the temperature at the cold junction. Therefore, often a temperature sensor, for example a NTC resistor, is integrated into thermopiles to measure the cold junction temperature (Figure 6.4.5). In order to achieve a maximum signal-to-noise ratio, the electronics for signal evaluation should also be integrated into the sensor housing (integrated signal processing). Figure 6.4.6 shows a typical evaluation circuit with a digital output. Figure 6.4.7 presents the model of a thermocouple for the electronic simulation of such circuits. In addition to thermoelectric voltage V_{th}, it also includes internal resistance R_{TP} of the thermocouples.

A complete sensor with thermopile, temperature sensor, signal processing, housing and – if required – optical components is commonly called thermopile module. It can be manufactured – according to application specifications – in a large variety of designs (Figure 6.4.8). Thermopile modules can be used as completely contact-free radiation thermometers. For imaging systems, it is possible to construct two-dimensional focal plane arrays (Figure 6.4.9).

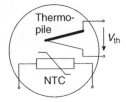

Figure 6.4.5   Schematic of a thermopile with integrated NTC resistor

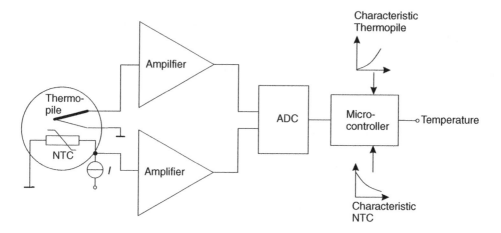

Figure 6.4.6   Typical schematic of a pyrometer with a thermopile with digital output [7]

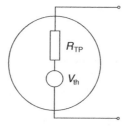

Figure 6.4.7   Electronic model of a thermopile

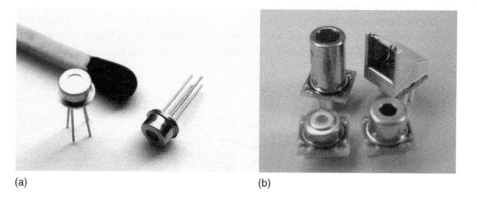

(a)   (b)

Figure 6.4.8   Miniaturised thermopile module, (a) sensor, (b) sensor with integrated signal processing
(By courtesy of Heimann Sensor GmbH, Dresden, Germany)

Figure 6.4.9   Monolithic 32×32 thermopile sensor array with digital output (By courtesy of Heimann Sensor GmbH, Dresden, Germany)

6.5   Pyroelectric Sensors

6.5.1   Principle

For a number of materials, particularly crystalline materials, positive and negative charges in a unit cell or a in a molecule do not coincide, but are spatially separated. This is called polarisation.[4] Polarisation can occur without any external impact (spontaneous polarisation) or be achieved by electric fields. Pyroelectric sensors utilise the temperature-dependence of the spontaneous polarisation of certain dielectric materials – so called pyroelectrics – for the thermoelectric coupling. Table 6.5.1 summarizes the classification of dielectrics. Pyroelectrics are piezoelectric crystals that have an electric dipole moment without any external electric field.

Figure 6.5.1 shows polarisation P as a function of field strength E for ferroelectrics. The function is not unambiguous, but shows hysteresis behaviour. The intersection of the curve with the ordinate is remnant polarisation P_R, which occurs in the crystal without any external electric field being applied.

Ferroelectrics are pyroelectrics with switchable polarisation, that is the direction of the spontaneous polarisation can be changed by an external electric field. This behaviour can be

Table 6.5.1   Classification of crystalline dielectric materials

Material Group				Example
Dielectrics				Tantalum pentoxide Ta_2O_5
	Piezoelectrics			Quartz SiO_2
		Pyroelectrics		Lithium tantalate $LiTaO_3$
			Ferroelectrics	Lead zirconate titanate PZT $(PbZr_xTi_{1-x}O_3)$

---

[4] Polarisation is the sum of all dipole moments per volume unit.

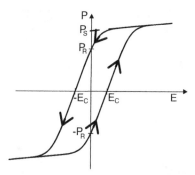

Figure 6.5.1 *P-E*-curve of ferroelectrics. The arrows show the direction in which we follow the curve when electric field strength E changes. P_S saturation polarisation; P_R remnant polarisation; E_C coercive field strength

used for the polarity of ferroelectrics. After applying large electric field strengths $\pm E$, that lead to saturation polarisation $\pm P_S$ and the return to $E = 0$, it is possible to achieve – depending on the polarity of E – the value $+P_R$ or $-P_R$ for the remaining polarisation.

As presented in Table 6.5.1, all pyroelectrics are also piezoelectrics. The HECKMANN diagram (Figure 6.5.2) illustrates the relations between thermal, mechanical and dielectric effects of dielectrics.

For piezoelectrics, spontaneous polarisation is changed by applying mechanical stress. Piezoelectrics, and consequently also pyroelectrics, show anisotropic material characteristics. In order to achieve a maximum responsivity, the direction of the spontaneous polarisation has to be in the same direction as the area normal of the sensor element.

As we can see in the HECKMANN diagram, both mechanical and thermal effects lead to changes in field strength. Thus, a change in temperature due to the pyroelectric effect directly causes a change in the dielectric displacement density (primary pyroelectric effect). A temperature change however also generates a change of mechanical stress T_m, which in turn leads to a change of displacement density due to piezoelectricity (secondary pyroelectric effect). In order to describe the individual effects, we will keep the state variables that are not analysed constant. Thus, it applies to the linear state equation of the displacement density that

$$\vec{D} = \frac{\delta \vec{D}}{\delta T}\bigg|_{T_m,E} \times \Delta T + \frac{\delta \vec{D}}{\delta T_m}\bigg|_{T,E} \times T_m + \frac{\delta \vec{D}}{\delta \vec{E}}\bigg|_{T,T_m} \times \vec{E} \qquad (6.5.1)$$

$\vec{D}$, T_m and $\vec{E}$ actually describe the changes Δ of these parameters. As the sensors considered here still operate in the linear range, we can use the variables themselves instead of their changes in the following. The indices in Equation 6.5.1 describe the state variables that are kept constant. All material parameters are tensors. The left-hand term of the sum in Equation 6.5.1 describes the pyroelectric effect. For pyroelectric coefficient π_P it applies for constant mechanical tension T_m and constant electric field strength E that

$$\vec{\pi}_P = \frac{\delta \vec{D}}{\delta T}\bigg|_{T_m,E} \qquad (6.5.2)$$

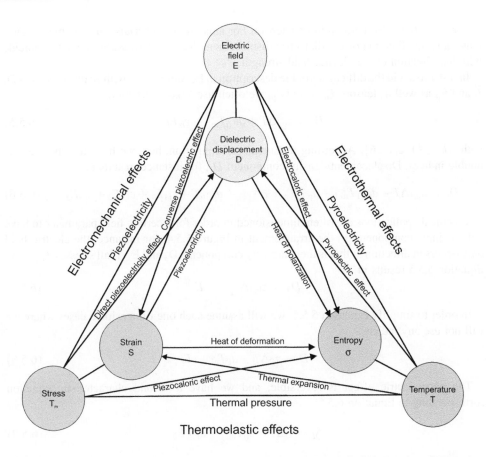

Figure 6.5.2 (caption text below)

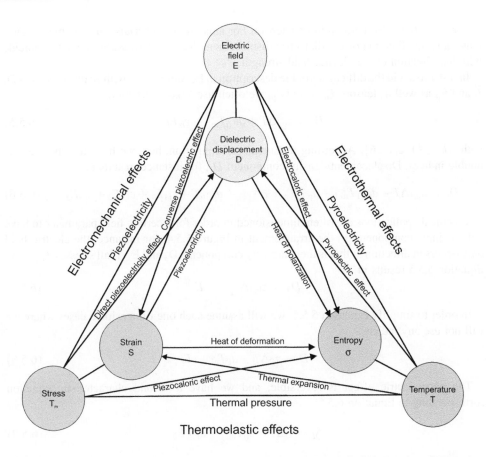

Figure 6.5.2 HECKMANN diagram [9] (By courtesy of DIAS Infrared GmbH, Dresden, Germany)

The middle term of the sum in Equation 6.5.1 describes the piezoelectric effect. For the piezoelectric coefficient, it applies for constant temperature T and constant electrical field strength E that

$$d_P = \left. \frac{\delta D}{\delta T_m} \right|_{T,E} \tag{6.5.3}$$

As both piezoelectric coefficient d_P and mechanical tension T_m are third- or second-order tensors, it is not possible any longer to provide a simple vectorial description as in Equation 6.5.2. The right-hand term of the sum in Equation 6.5.1 describes the dielectricity. For permittivity, it applies for constant temperature T and constant mechanical tension T_m that

$$\varepsilon = \left. \frac{\delta \vec{D}}{\delta \vec{E}_m} \right|_{T,T_m} \tag{6.5.4}$$

Permittivity is also a second-order tensor. For anisotropic materials, such as pyroelectric materials, the direction of the dielectric displacement does not necessarily have to coincide with the direction of the electric field strength.

In order to avoid the difficulties of the description in Equation 6.5.1 with scalar T, vectors $\vec{D}$, $\vec{E}$ and $\vec{\pi}_p$ as well as tensors T_m, ε and d_p, we often use VOIGT'S notation:

$$D_i = \pi_{p,i}\Delta T + d_{P,ij}T_{m,j} + \varepsilon_{ik}E_k \qquad (6.5.5)$$

with $i,k \in \{1;2;\ldots6\}$. According to EINSTEIN'S summation, here we have to sum over the double indices. Displacement density component D_1, for instance, results in

$$D_1 = \pi_{p,1}\Delta T + d_{p,11}T_{m,1} + d_{p,12}T_{m,2} + \cdots + d_{p,16}T + \varepsilon_{11}E_1 + \varepsilon_{12}E_2 + \varepsilon_{13}E_3 \qquad (6.5.6)$$

Technical applications are often dimensioned in order for a certain field parameter to have only a single component. In the arrangement in Figure 6.5.4, for instance, the electric field only occurs in direction 3, that is there are only components D_3 and E_3. If all stresses $T_{m,j} = 0$, Equation 6.5.5 results in

$$D_3 = \pi_{P,3}\Delta T + \varepsilon_{33}E_3 \qquad (6.5.7)$$

In order to simplify Equation 6.5.5, we will assume such one-dimensional cases where we will not use any indices:

$$D = \pi_P\Delta T + d_P T_m + \varepsilon E \qquad (6.5.8)$$

The piezoelectric effect is reversible, and we have a second characteristic equation corresponding to Equation 6.5.5:

$$S_l = \alpha_l\Delta T + s_{lj}T_{m \cdot j} + d_{P,lk}E_k \qquad (6.5.9)$$

Here, S_l is normal and shear strain ($l = 1$–3 and 4–6, respectively), α_l the coefficient of linear thermal expansion, and s_{lj} the moduli of elasticity (inverse E-moduli). In the one-dimensional case, it becomes

$$S = \alpha \, \Delta T + s \, T_m + d_p E \qquad (6.5.10)$$

Pyroelectric sensors now measure the field strength change (voltage mode at $D=0$) or the dielectric displacement change (current mode at $E=0$). It is not possible, though, to determine the origin (temperature change ΔT or a change of mechanical stress ΔT_m) of the respective signal change. Mainly the accelerations that affect the sensor can lead to a change in mechanical stress that in turn generates a sensor output signal. This effect is called microphony. For pyroelectric radiation sensors, this microphony or acceleration responsivity is often very disturbing as during operation the sensors are often exposed to shocks. The piezoelectric microphony signal is an interference signal that can become considerably stronger than the useful signal.

Usually, we look at both effects separately und neglect their coupling, that is for the description of the responsivity in relation to the radiant flux we assume a constant mechanical stress T_m or, respectively, for the description of the microphony, sensor temperature T_S is supposed to be constant.

Table 6.5.2 Parameters of pyroelectric materials [10]

Type	Material	Pyroelectric coefficient π_P [$\mu C/(m^2\,K)$]	Relative permittivity ε_r	Loss factor $\tan\delta$	Specific heat capacity c_p [$J/(m^3 K)$]	CURIE temperature T_C [K]
Monocrystals	TGS^a	280	27	1×10^{-2}	2.30×10^{-6}	49
	DTGS^a	550	18	2×10^{-2}	2.30×10^{-6}	61
	LiTaO$_3$	170	47	1×10^{-3}	3.20×10^{-6}	603
	LiNbO$_3$	80	30	5×10^{-4}	2.90×10^{-6}	1480
Ceramics	PZT	400	290	3×10^{-3}	2.50×10^{-6}	230
	BST	7000	8800	4×10^{-3}	2.55×10^{-6}	25
	PST	3500	2000	5×10^{-3}	2.70×10^{-6}	25
Polymer	PVDF	27	12	1×10^{-2}	2.40×10^{-6}	80

^{a}Alanine-doped.

Pyroelectric materials can be divided into single crystals, ceramics and ferroelectric polymers (Table 6.5.2).

Triglycinsulfate (TGS), deuterated triglycinsulfate (DTGS), lithium niobate (LiNbO$_3$) and lithium tantalate (LiTaO$_3$) belong to single-crystal pyroelectrics. Ferroelectric TGS and its modifications, such as alanine-doped, deuterated TGS (DTGS: LA), show excellent pyroelectric characteristics, but are technologically difficult to handle (long crystal-growing periods; low CURIE temperature T_C, above which the pyroelectric characteristics disappear; very fragile; hygroscopic). Therefore, they are only used in very expensive sensors that require a particularly large signal-to-noise ratio. Lithium tantalate and lithium niobate have a sufficiently high CURIE temperature T_C and consequently, for the operating temperature range of 0–70 °C that is relevant for sensors, they show an insignificant dependence of the pyroelectric coefficient on the temperature of only a few percent. Lithium tantalate, above all, is considered the standard material of pyroelectric sensors.

There is a large number of ferroelectric ceramics. For use in pyroelectric sensors the following materials are important: ferroelectric oxide ceramics lead zirconate titanate (PZT or PbZr$_{1-x}$Ti$_x$O$_3$) in different compositions of x, barium strontium titanate (BST) and lead scandium tantalate (PST). They all have in common the perovskite structure of the crystal structure which is the precondition of spontaneous polarisation [11]. Also, lithium tantalate and lithium niobate show perovskite-like crystal structures. Ceramics are polycrystalline. During manufacturing, areas of identical polarisation – the so-called domains – emerge in the ceramics as well as in monocrystals. By poling the domains get the same direction. Inexpensive sensors mainly use ferroelectric ceramics.

The most commonly known representative of ferroelectric polymers is polyvinylidene fluoride (PVDF). The polymer consists of a chain based on the monomer CF$_2$-CH$_2$. PVDF shows an amorphous matrix with imbedded crystallites. It has to be poled by an electric field that exceeds coercive field strength E_C of approximately 100 MV/m. PVDF can be manufactured as a thin film with a thickness of only a few micrometres. By straining, it is transferred into a ferroelectric modification. PVDF is considered an inexpensive pyroelectric material.

In addition to the pyroelectric effect, we can use capacity or impedance measurements for dielectric materials to utilise the dependence of relative permittivity on temperature. Such sensors are called dielectric bolometers. It is also possible to apply an external electric field to cause a polarisation or to enlarge an existing spontaneous polarisation, which then also is dependent on temperature. In this case, we talk about induced pyroelectricity.

6.5.1.1 Responsivity of Pyroelectric Sensors

The description of the pyroelectric effect requires the condition of a constant mechanical stress T_m. In general, it applies for the dielectric displacement that

$$\vec{D} = \varepsilon_0 \vec{E} + \vec{P} \qquad (6.5.11)$$

The dielectric displacement depends on the direction. In the following, the modulus of the dielectric displacement is always considered in the direction of the area normal of the sensor element, as it is the only one to generate a sensor signal. For polarisation P, the following basic equation applies:

$$P = \chi \varepsilon_0 E + P_S \qquad (6.5.12)$$

or

$$P = \chi \varepsilon_0 E + P_R + d_P T_m + \pi_P \Delta T \qquad (6.5.13)$$

with susceptibility χ, spontaneous polarisation P_S and remnant polarisation P_R. It applies for the susceptibility that:

$$\chi = \varepsilon_r - 1 \qquad (6.5.14)$$

For dielectric materials, due to $\chi \geq 0$ it always applies that $\varepsilon_r \geq 1$. In pyroelectric materials, the spontaneous polarisation always depends on temperature (Figure 6.5.3a). Above CURIE temperature T_C, it is zero. The dielectric material is then in the paraelectric state.

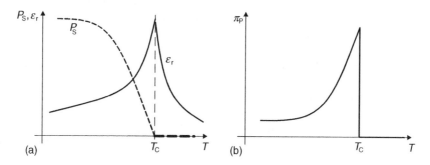

Figure 6.5.3 Temperature dependence (a) of spontaneous polarisation P_S and relative permittivity ε_r as well as (b) of the pyroelectric coefficient π_P [12]

The temperature dependence of the spontaneous polarisation is called pyroelectricity. For pyroelectric coefficient π_P, it thus applies according to Equation 6.5.2:

$$\pi_P = \frac{dP}{dT} \qquad (6.5.15)$$

The unit of pyroelectric coefficient π_P is $C/(m^2 K)$. Figure 6.5.3b illustrates that π_P is heavily dependent on the temperature, at least close to CURIE temperature T_C.

It results from the dielectric displacement in Equation 6.5.8 that

$$D = \varepsilon_0 \varepsilon_r E + P_R + d_P T_M + \pi_P \Delta T_S \qquad (6.5.16)$$

Due to the dipole characteristics of pyroelectric materials, there are electric charges on the sensor surface that can be compensated by freely moving electrons in the environment. Surface area charge density σ_Q is equal to the dielectric displacement:

$$\sigma_Q = D = \frac{Q}{A_S} \qquad (6.5.17)$$

If the dielectric displacement now changes, the surface charge also changes. If we deposit electrodes onto the surface of a pyroelectric element with electrodes, we can measure the changes in surface charge (Figure 6.5.4). The overlap of front and back electrode corresponds to the actual detector element with area A_S and thickness d_S. The change in surface charge induces a (displacement) current:

$$I_Q(t) = A_S \frac{dD}{dt} = \varepsilon_0 \varepsilon_r A_S \frac{dE}{dt} + \pi_P A_S \frac{d\Delta T_S}{dt} \qquad (6.5.18)$$

As the charges are quasistationary in dielectrics, in this case current flow causes the deflection of charges around their resting positions. The pyroelectric detector element constitutes a plate capacitor. This means it applies that

$$E = \frac{V_Q}{d_S} \qquad (6.5.19)$$

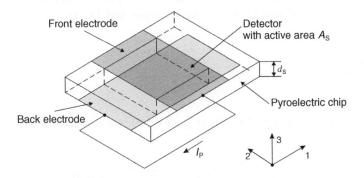

Figure 6.5.4   Set-up of a pyroelectric detector

With Equations 6.5.18 and 6.5.19 we arrive at

$$I_Q(t) = \varepsilon_0 \varepsilon_r \frac{A_S}{d_S} \frac{dV_Q}{dt} + \pi_P A_S \frac{d\,\Delta T_S}{d\,t} \tag{6.5.20}$$

For a harmonic excitation with a modulation or chopper frequency ω_{Ch} in the frequency domain, Equation 6.5.20 results in

$$\underline{I}_Q(\omega_{Ch}) = \varepsilon_0 \varepsilon_r \frac{A_S}{d_S} j\omega_{Ch}\underline{V}_Q + \pi_P A_S\, j\omega_{Ch}\underline{\Delta T}_S \tag{6.5.21}$$

If we short-circuit the electrodes, the voltage becomes $\underline{V}_Q = 0$. This means that also the external electric field becomes zero ($E = 0$). This operating mode is called current mode. It applies for pyroelectric short-circuit current $\underline{I}_P$ that

$$\underline{I}_P = j\omega_{Ch}\pi_P A_S \underline{\Delta T}_S \tag{6.5.22}$$

Pyroelectric current $\underline{I}_P$ is directly proportional to temperature change $\underline{\Delta T}_S$ and chopper frequency ω_{Ch}. However, if we leave the electrodes open, we talk about voltage mode ($\underline{I}_Q = 0$). In this case, the dielectric displacement is zero ($D = 0$) and the generated charges induce voltage $\underline{V}_Q$ at the electrodes:

$$\underline{V}_Q = -\frac{\pi_P}{\varepsilon_0 \varepsilon_r} d_S \underline{\Delta T}_S = -\frac{\pi_P}{C_P} A_S \underline{\Delta T}_S \tag{6.5.23}$$

The induced voltage $\underline{V}_Q$ is directly proportional to temperature change $\underline{\Delta T}_S$ and independent of the frequency.

Figure 6.5.4 presents the electric model of a pyroelectric sensor. Current source $\underline{I}_P$ characterises the short-circuit caused according to Equation 6.5.22 by temperature change $\underline{\Delta T}_S$. The sensor element is a plate capacitor that consists of capacity C_P, DC resistance R_P and dielectric loss resistance $R_{\tan\delta}$. As pyroelectrics are good insulators ($R_P > 10^{12}\ \Omega$) it is possible to neglect the DC resistance in most cases. Often we consider loss angle $\tan\delta$ as the ratio of AC resistance $1/\omega C_p$ of the capacitor and DC resistance $R_{\tan\delta}$. Thus, it results that

$$R_{\tan\delta} = \frac{1}{\omega_{Ch} C_P \tan\delta}. \tag{6.5.24}$$

Using the electric small-signal model (Figure 6.5.5), we can calculate the pyroelectric voltage as

$$\underline{V}_P = \frac{R'_P}{1 + j\omega_{Ch}R'_P C_P}\underline{I}_P = \frac{j\omega_{Ch}R'_P}{1 + j\omega_{Ch}\,\tau_E}\pi_P A_S \underline{\Delta T}_S \tag{6.5.25}$$

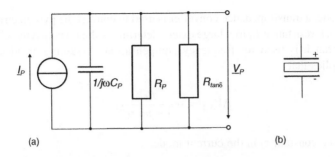

Figure 6.5.5 (a) Electric small-signal model of a pyroelectric detector and (b) electric circuit symbol

with resistance R'_P according to

$$\frac{1}{R'_P} = \frac{1}{R_P} + \frac{1}{R_{\tan\delta}}$$ (6.5.26)

and electrical time constant τ_E

$$\tau_E = R'_P C_P$$ (6.5.27)

Using Equation 6.5.25, we arrive at Equation 6.5.23 by neglecting loss resistance R'_P ($R'_P \to \infty$) and taking into account that voltage $\underline{V}_Q$ is defined in direction of the pyroelectric current:

$$\underline{V}_P = -\underline{V}_Q$$ (6.5.28)

Figure 6.5.6 presents the basic circuits of a pyroelectric sensor in current and voltage mode.

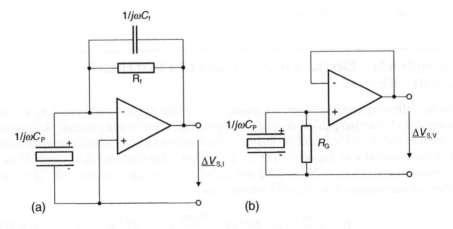

Figure 6.5.6 Basic circuit of a pyroelectric detector element (a) in current mode and (b) in voltage mode

In current mode, a transimpedance converter is used to convert the pyroelectric current into voltage. Feedback resistance R_f to a large extent determines the responsivity. Capacitor C_f in the feedback branch is used for frequency compensation. Assuming an ideal operational amplifier, it applies that

$$\underline{\Delta V}_{S,I} = -\frac{R_f}{1 + j\omega \tau_I} \underline{I}_P \qquad (6.5.29)$$

with electric time constant τ_I in the current mode:

$$\tau_I = R_f C_f \qquad (6.5.30)$$

In voltage mode, an impedance converter (voltage follower) is used to provide at the output a low-impedance voltage over the sensor element. Series resistance R_G determines the operating point of the impedance converter. Assuming an ideal operational amplifier (OV) with gain a_v, it applies that:

$$\underline{\Delta V}_{S,V} = \frac{R_V a_v}{1 + j\omega \tau_V} \underline{I}_P \qquad (6.5.31)$$

with electric time constant τ_V in the voltage mode

$$\tau_V = R_V C_V \qquad (6.5.32)$$

as well as resistance R_V and capacity C_V in the input circuit according to

$$\frac{1}{R_V} = \frac{1}{R_P} + \frac{1}{R_{\tan\delta}} + \frac{1}{R_G} \qquad (6.5.33)$$

and

$$C_V = C_P + C_E \qquad (6.5.34)$$

with input capacity C_E of the impedance converter.

Example 6.9: Electric Characteristics of a Pyroelectric Detector Element

For the following considerations, we assume an electric small-signal behaviour which means that all material parameters can be supposed to be linear and constant. A detector element of size $A_S = 1 \times 1 \, \text{mm}^2$ is situated in the middle of a pyroelectric chip consisting of lithium tantalate with a dimension of $2 \times 2 \, \text{mm}^2$ and a chip thickness d_S of $10 \, \mu\text{m}$. Using the material parameters in Table 6.5.2 and resistivity ρ of approximately $10^{12} \, \Omega\text{m}$, the following parameters of the detector element result:

$$R_P = \rho \frac{d_S}{A_S} = 1 \times 10^{12} \, \Omega\text{m} \frac{10 \, \mu\text{m}}{1 \, \text{mm}^2} = 1 \times 10^{13} \, \Omega \qquad (6.5.35)$$

$$C_P = \frac{\varepsilon_0 \varepsilon_r A_S}{d_S} = \frac{8.854 \times 10^{-12}\, F\, m^{-1} \times 47 \times 1\, mm^2}{10\, \mu m} = 41.6\, pF \qquad (6.5.36)$$

$$R_{\tan\delta}(f_{Ch} = 10\, Hz) = \frac{1}{\omega_{Ch}\, C_P\, \tan\delta} = \frac{1}{62.83\, Hz \times 41.6\, pF \times 1 \times 10^{-3}} = 3.83 \times 10^{11}\, \Omega \qquad (6.5.37)$$

Using the calculated values, it applies that

$$R'_P = R_{\tan\delta} = 3.83 \times 10^{11}\, \Omega$$

For a detector temperature change of $\Delta\tilde{T}(f_{Ch} = 10\, Hz) = 0.1\, K$, for instance, the effective value of the pyroelectric current is

$$\tilde{i}_P(f_{Ch} = 10\, Hz) = \omega_{Ch}\, \pi_P A_S \Delta\tilde{T}_S = 62.83\, Hz \times 230 \times 10^{-6}\, Cm^2\, K^{-1}$$
$$\times 1\, mm^2 \times 0.1\, K = 1.45\, nA \qquad (6.5.38)$$

For the open circuit of the detector element, the resulting voltage is

$$\tilde{v}_P(f_{Ch} = 10\, Hz) = \frac{R'_P\, \tilde{i}_P}{\sqrt{1 + (\omega_{Ch} R'_P C_P)^2}}$$

$$= \frac{3.83 \times 10^{11}\, \Omega \times 1.45\, nA}{\sqrt{1 + (62.83\, Hz \times 7.65 \times 10^{11}\, \Omega \times 41.6\, pF)^2}} = 0.28\, V \qquad (6.5.39)$$

Without electrical wiring, for instance during the manufacturing process, the open-circuit voltage over the detector element can become very large. Temperature changes of only a few $10\, K$ can cause voltages between the sensor electrodes that are above the material's breakdown voltage and lead to flashovers between adjacent electrodes.

According to Equations 6.2.1, 6.1.19 and 6.2.17, heat capacity C_{th}, thermal conductance $G_{th,R}$ due to heat radiation and thermal time constant τ_{th} amount to

$$C_{th} = c_P A_S d_S = 3.2 \times 10^6\, Ws\, m^{-3}\, K^{-1} \times 1\, mm^2 \times 10\, \mu m = 3.2 \times 10^{-5}\, Ws/K \qquad (6.5.40)$$

$$G_{th,S} = 4\varepsilon\sigma T_S^3 A_S = 4 \cdot 1 \cdot 5.671 \times 10^{-8}\, Wm^{-2}\, K^{-4} \times (300\, K)^3 \times 1\, mm^2$$
$$= 6.12 \times 10^{-6}\, W/K \qquad (6.5.41)$$

$$\tau_{th} = \frac{C_{th}}{G_{th}} = \frac{3.2 \times 10^{-5}\, Ws/K}{6.12 \times 10^{-6}\, W/K} = 5.2\, s \qquad (6.5.42)$$

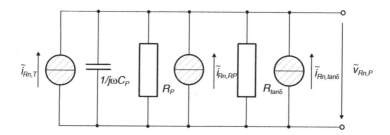

Figure 6.5.7 Noise model of a pyroelectric detector element

Example 6.10:   Noise Current of a Pyroelectric Detector Element

Figure 6.5.5a can be used to directly derive the noise model of a pyroelectric detector element (Figure 6.5.7).

OHMIC resistance R_P and $R_{\tan\delta}$ each contribute to total noise. The noise current of DC resistance $\tilde{u}_{R,R}$ is calculated from Equation 4.2.3 and $\tan\delta$ noise $\tilde{i}_{Rn,\tan\delta}$ from Equation 4.2.7. The only non-electric noise source is temperature fluctuation noise current $\tilde{i}_{Rn,T}$ (Section 4.2.5) that is calculated from Equations 4.2.45 and 6.5.22:

$$\tilde{i}_{Rn,T}^2 = \omega_{Ch}^2 \pi_P^2 A_S^2 \frac{4k_B}{G_{th}} \frac{T_S^2}{1+\omega_{Ch}^2 \tau_{th}^2} \tag{6.5.43}$$

Without any restrictions of the noise bandwidth and neglecting the temperature fluctuation noise, according to Equation 4.2.10, the following kTC noise applies to the sensor element

$$\tilde{v}_{R,kTC} = \sqrt{\frac{k_B T_S}{C_P}} = \sqrt{\frac{1.381 \times 10^{-23} \,\text{Ws}\,\text{K}^{-1} \times 300\,\text{K}}{41.6\,\text{pF}}} = 10\,\mu\text{V} \tag{6.5.44}$$

The total noise current of the sensor element corresponds to:

$$\tilde{i}_{rn,P}^2 = \tilde{i}_{Rn,T}^2 + \tilde{i}_{Rn,RP}^2 + \tilde{i}_{Rn,\tan\delta}^2 \tag{6.5.45}$$

Example 6.11:   Noise in Current Mode

In current mode, the detector element described in Example 6.9 is connected as presented in Figure 6.5.6a. An operational amplifier with a small input current (FET input stage) and very low current noise is suitable as an amplifying element. In the example, the current noise is $\tilde{i}_{Rn,OV} = 0.5\,\text{fA}/\sqrt{\text{Hz}}$ and the voltage noise $\tilde{v}_{Rn,OV}(100\,\text{Hz}) = 30\,\text{nV}/\sqrt{\text{Hz}}$, with

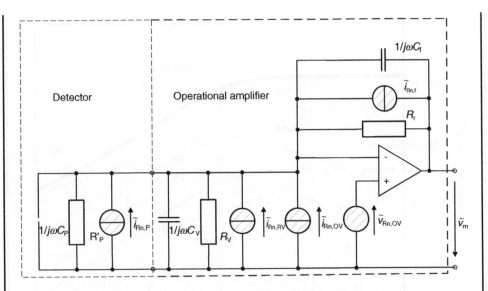

Figure 6.5.8   Noise model of a pyroelectric detector element in current mode

the voltage noise showing a typical $1/f$-behaviour. Figure 6.5.8 shows the noise model of a pyroelectric sensor element in current mode.

The impedance of the circuit results from the impedance of the feedback branch of the operational amplifier:

$$Z_\text{f} = \frac{R_\text{f}}{\sqrt{1 + \omega^2 \tau_I^2}} \tag{6.5.46}$$

with electric time constant τ_I of

$$\tau_I = R_\text{f} C_\text{f} = 1 \times 10^{10} \Omega \times 2\,\text{pF} = 20\,\text{ms}$$

Noise voltage $\tilde{v}_\text{Rn}$ at the output of the connected pyroelectric detector in current mode is calculated as

$$\tilde{v}_\text{Rn}^2 = (\tilde{i}_\text{Rn,P}^2 + \tilde{i}_\text{Rn,RV}^2 + \tilde{i}_\text{Rn,OV}^2 + \tilde{i}_\text{Rn,f}^2)Z_\text{f}^2 + \left(1 + \frac{Z_\text{f}}{Z_\text{P}}\right)^2 \tilde{v}_\text{Rn,OV}^2 \tag{6.5.47}$$

with

$$Z_\text{P} = \frac{R'_\text{P} \| R_V}{\sqrt{1 + \omega^2 (C_\text{P} + C_V)^2 (R'_\text{P} \| R_V)^2}} \tag{6.5.48}$$

with $1 + \dfrac{Z_\text{f}}{Z_\text{P}} \approx 1$ applying for this example. Figure 6.5.9 presents the total noise voltage according to Equation 6.5.47 and the individual noise contributions.

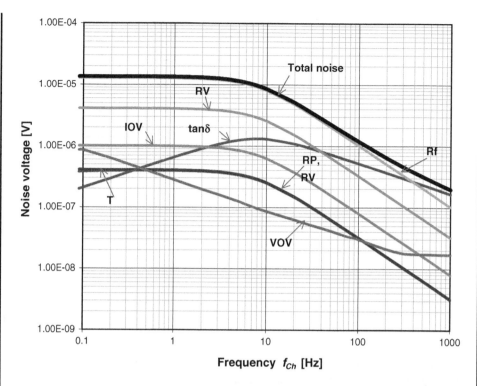

Figure 6.5.9 Normalised voltage $\tilde{v}_{Rn}$ of a pyroelectric detector element in current mode; for the used component parameters of the LiTaO$_3$ sensor, see Table 6.5.3; for abbreviations regarding the noise components, see Table 6.5.4

The value of feedback resistance R_f determines the relation of the individual noise contributions. Above all, feedback resistance R_f and consequently the gain of the operational amplifier have to be set at large values in order for the voltage noise of the amplifier not to be dominant.

Table 6.5.3 Component parameters of the circuit in Figure 6.5.7

Denomination	Symbol	Value
Loss resistance of the pyroelectric element	tanδ	0.001
DC resistance of the pyroelectric element	R_P	$1 \times 10^{-13}\,\Omega$
Input resistance of the OV	R_V	$1 \times 10^{11}\,\Omega$
Feedback resistance	R_f	$1 \times 10^{10}\,\Omega$
Feedback capacity	C_f	2 pF
Sensor temperature	T_S	300 K

Table 6.5.4   Abbreviations in Figures 6.5.9 and 6.5.11

Noise voltage component	Abbreviation	Calculation rule for Figure 6.5.8	Calculation rule for Figure 6.5.10
Temperature fluctuation noise	T	$Z_f\,\tilde{i}_{Rn,T}$	$Z_V\tilde{i}_{Rn,T}$
Resistance noise of the pyroelectric element	Rp	$Z_f\tilde{i}_{Rn,RP}$	$Z_V\tilde{i}_{Rn,RP}$
Noise of the loss resistance	tanδ	$Z_f\tilde{i}_{Rn,\tan\delta}$	$Z_V\,\tilde{i}_{Rn,\tan\delta}$
Noise of the input resistance of the OV	RV	$Z_f\,\tilde{i}_{Rn,RV}$	—
Noise of the feedback resistance	Rf	$Z_f\tilde{i}_{Rn,f}$	—
Series resistance	RG	—	$Z_V\tilde{i}_{Rn,RG}$
Current noise of the amplifier	IOV	$Z_f\tilde{i}_{Rn,OV}$	$Z_V\tilde{i}_{Rn,OV}$
Voltage noise of the amplifier	VOV	$\tilde{v}_{Rn,OV}$	$\tilde{v}_{Rn,OV}$
Noise voltage at the output	Total noise	Equation 6.5.47	Equation 6.5.49

Example 6.12:   Noise in Voltage Mode

The detector element described in Example 6.9 is connected according to Figure 6.5.6b. A voltage follower with a very small input current and very low voltage noise is a suitable amplifier device. In the example, it applies for noise current $\tilde{i}_{Rn,OV} = 5\,\text{fA}/\sqrt{\text{Hz}}$ and noise voltage $\tilde{v}_{Rn,OV}(100\,\text{Hz}) = 5\,\text{nV}/\sqrt{\text{Hz}}$, with the noise voltage showing a typical $1/f$ behaviour. In the noise model (Figure 6.5.10), the input resistance of the amplifier was neglected as it is substantially larger than R_G.

The noise output voltage in voltage mode is

$$\tilde{v}_{Rn}^2 = [(\tilde{i}_{Rn,P}^2 + \tilde{i}_{Rn,RG}^2 + \tilde{i}_{Rn,OV}^2)Z_V^2 + \tilde{v}_{Rn,OV}^2]a_V^2 \tag{6.5.49}$$

with impedance Z_V of the input circuit (Figure 5.6.11)

$$Z_V = \frac{R_P'||R_G}{\sqrt{1+\omega^2\,\tau_V^2}} \tag{6.5.50}$$

and gain of the amplifier

$$a_V = 1 \tag{6.5.51}$$

Electric time constant τ_V is

$$\tau_V = (C_P + C_V)R_P'||R_G \tag{6.5.52}$$

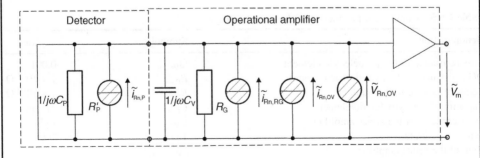

Figure 6.5.10   Noise model of a pyroelectric sensor element in voltage mode

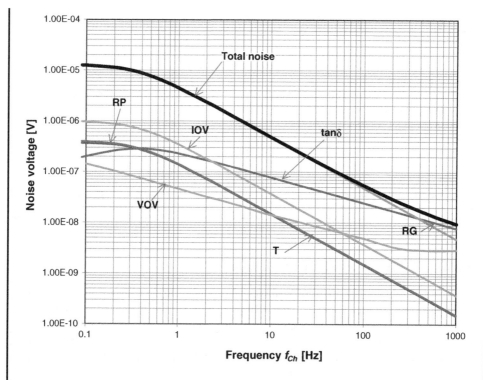

Figure 6.5.11 Normalised noise voltage $\tilde{v}_{Rn}$ of a pyroelectric sensor element in voltage mode; for the used component parameters of the LiTaO$_3$ sensor, see Table 6.5.5; for abbreviations regarding the noise components, see Table 6.5.4

6.5.1.2 Microphony

In order to describe microphony, we will define the sensor parameter acceleration responsivity R_{Va}:

$$R_{Va} = \frac{\tilde{V}_m}{\tilde{a}} \tag{6.5.53}$$

Table 6.5.5 Component parameters of the circuit in Figure 6.5.9

Denomination	Symbol	Value
Loss resistance of the pyroelectric element	tanδ	0.001
DC resistance of the pyroelectric element	R_P	1×10^{13} Ω
Series resistance in the input circuit	R_G	1×10^{10} Ω
Capacity of the pyroelectric element	C_P	41.6 pF
Input capacity of the series amplifier	C_V	1 pF
Sensor temperature	T_S	300 K
Gain of the series amplifier	a_V	1

The acceleration responsivity is usually related to the acceleration due to gravity

$$\tilde{a} = g = 9.80665 \text{ m s}^{-2} \tag{6.5.54}$$

The unit is V/g.

For pyroelectric sensors, we are mainly interested in the piezoelectric effect in the direction of the area normal of the sensor element. This way, we can use simple, one-dimensional formulas. Thus, it results from Equation 6.5.8

$$D = d_P T_m + \varepsilon_0 \varepsilon_r E \tag{6.5.55}$$

with material constants d_P and ε_r in direction of the area normal of the sensor element. Equation 6.5.52 can be used to calculate the electrical field strength of a sensor element in voltage mode ($D = 0$):

$$E = -\frac{d_P}{\varepsilon_0 \varepsilon_r} T_m \tag{6.5.56}$$

The respective voltage V_m results by integrating over the sensor thickness:

$$V_m = -\int_0^{d_S} E(z) dz \tag{6.5.57}$$

If the sensor element is now sinusoidally accelerated in z-direction, the sensor element is affected by force $\underline{F}$:

$$\underline{F} = \rho A_S z \underline{a} \tag{6.5.58}$$

with density ρ and acceleration $\underline{a}$. Mechanical stress $\underline{T}_m$ in the sensor element is calculated as

$$\underline{T}_m = \frac{F}{A_S} = \rho z \underline{a} \tag{6.5.59}$$

In voltage mode, using Equations 6.5.56–6.5.59, piezoelectric voltage $\underline{V}_m$ becomes:

$$\underline{V}_m = \int_0^{d_S} \frac{d_P}{\varepsilon_0 \varepsilon_r} z \, \rho \underline{a} dz = \frac{1}{2} \frac{d_P}{\varepsilon_0 \varepsilon_r} \rho d_S^2 \underline{a} \tag{6.5.60}$$

The acceleration responsivity is then according to Equation 6.5.53

$$R_{Va} = \frac{|\underline{V}_m|}{|\underline{a}|} = \frac{1}{2} \frac{d_P}{\varepsilon_0 \varepsilon_r} \rho d_S^2 \tag{6.5.61}$$

Example 6.13: Acceleration Responsivity of Pyroelectric Sensors

The acceleration responsivity in Equation 6.5.61 describes microphony for forces in polarisation direction (longitudinal effect). In practice, there may also occur forces that are transverse to the polarisation (transversal effect) and shear forces (shearing effect). The voltage relations in the detector element mainly depend on the actual design of the sensor. In this context, the structural design and, in particular, the suspension of the sensor chip are important. In practice, it is only possible to calculate the acceleration responsivity using numerical models, such as finite-element methods (FEM) or SPICE [13–15]. Figure 6.5.12

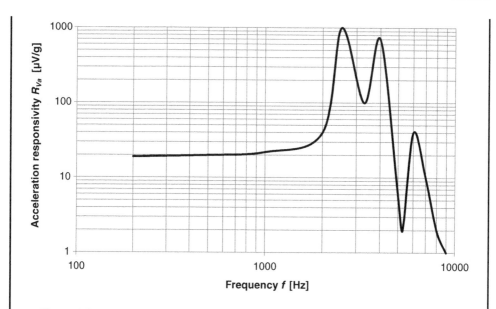

Figure 6.5.12   Simulated acceleration responsivity of a pyroelectric LiTaO₃ sensor [13]

shows the principal curve of the acceleration responsivity of the type of sensor shown in
Figure 6.5.16. The large fluctuations of the acceleration responsivity at a few kilohertz are
caused by the sensor chip's mechanical resonances.

6.5.2   Thermal Resolution

The voltage responsivity of a pyroelectric sensor is calculated applying Equations 6.5.22
and 6.5.29 or, respectively, Equations 6.2.23 and 6.5.31 as well as Equation 6.2.27:

$$R_V = \frac{\alpha \, \pi_P \, T_R \, A_S}{G_{th}} \frac{\omega_{Ch}}{\sqrt{1 + \omega_{Ch}^2 \, \tau_{th}^2}} Z_E \tag{6.5.62}$$

Impedance Z_E results from the input impedance of the sensor connection and – for the
voltage mode – becomes according to Equations 6.5.50 and 6.5.51

$$Z_E = a_V \, Z_V \tag{6.5.63}$$

and for the current mode according to Equation 6.5.47

$$Z_E = Z_f \tag{6.5.64}$$

For a constant radiation ($\omega_{Ch}=0$), responsivity R_V becomes zero (Figure 5.1.2b).
Pyroelectric sensors are not constant-light-sensitive! The reason is that according to
Equation 6.5.22, pyroelectric short-circuit current I_P depends on time change $d(\Delta T_S)/dt$.
In other words, the temperature change that is caused in the pyroelectric elements is in turn
compensated by resistances R_P and $R_{\tan\delta}$ (Figure 6.5.4). From this, it results that pyroelectric

sensors have to be modulated, that is choppered, for applications with constant or only slowly changing radiation intensities.

For sufficiently large chopper frequencies $(\omega_{Ch}^2 \gg 1/\tau_{th}^2)$, it applies for responsivity R_V that

$$R_V = \frac{\alpha \, \pi_P \, T_R \, A_S}{C_{th}} Z_E = \frac{\alpha \, \pi_P \, T_R}{c_S \, d_S} Z_E \qquad (6.5.65)$$

If we choose a chopper frequency that is sufficiently large regarding the electric time constant $(\omega_{Ch}^2 \gg 1/\tau_E^2)$, Equation 6.5.65 becomes

$$R_V = \frac{\alpha \pi_P T_R}{c_S d_S} \frac{R_E a_v}{\omega_{Ch}} \qquad (6.5.66)$$

It then applies for the voltage mode (Figure 6.5.5b) that

$$R_E = R_G \qquad (6.5.67)$$

or for the current mode (Figure 6.5.5a)

$$R_E = R_f \qquad (6.5.68)$$

For the current mode, it follows from Equation 6.5.65 the often stated equation for the responsivity of pyroelectric sensors

$$R_V = \frac{\alpha \, \pi_P \, T_R}{c_S \, d_S} \frac{1}{\omega_{Ch} C_f} \qquad (6.5.69)$$

and for the voltage mode – with $C_V \approx C_P$, in addition – the equation

$$R_V = \frac{\alpha \pi_P T_R}{c_S d_S} \frac{a_v}{\omega_{Ch} C_P} = \frac{\alpha \, \pi_P \, T_R}{c_S \, \varepsilon_0 \, \varepsilon_r} \frac{a_v}{\omega_{Ch} \, A_S} \qquad (6.5.70)$$

The responsivity is indirectly proportional to chopper frequency ω_{Ch}.
For the current mode, also the case $\omega_{Ch}^2 R_f^2 C_f^2 \ll 1$ is of interest:

$$R_V = \frac{\alpha \, \pi_P \, T_R \, R_f}{c_S \, d_S} \qquad (6.5.71)$$

The responsivity is then independent of the chopper frequency.

Example 6.14:   Responsivity of a Pyroelectric Sensor

We will calculate the current- and voltage-mode responsivity of the sensor in Examples 6.5.1–6.5.4. Figure 6.5.13 shows the voltage responsivity in current and voltage mode.

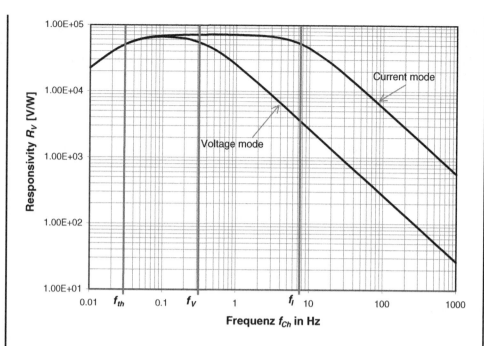

Figure 6.5.13 Responsivity R_V of a pyroelectric sensor in current and voltage mode versus chopper frequency

As thermal conductance, we apply the thermal conductance through radiation according to Equation 6.5.41. Thermal time constant τ_{th} determines the lower frequency limit:

$$f_{th} = \frac{1}{2\pi\tau_{th}} = \frac{1}{2\pi \times 5.23\,\text{s}} = 0.03\,\text{Hz} \tag{6.5.72}$$

As in addition to the thermal conductance through radiation, there is heat conduction through the ambient gas (air) and the chip mountings, realistic thermal frequency limits are in the range of 0.3 and 0.9 Hz. The upper frequency limit of responsivity R_V is calculated using the electric time constant. It applies to current mode that

$$f_I = \frac{1}{2\pi\tau_I} = \frac{1}{2\pi \times 0.02\,\text{s}} = 7.96\,\text{Hz} \tag{6.5.73}$$

For the electric time constant in voltage mode, it follows with $R'_P = R_{tan\delta}$ applying Equation 6.5.52 and $\omega_{Ch} = 10\,\text{Hz}$:

$$\tau_V = \frac{R_G(C_P + C_V)}{1 + \omega_{Ch}\,C_P\,R_G\tan\delta} = \frac{1 \times 10^{10}\,\Omega \times (41.6\,\text{pF} + 1\,\text{pF})}{1 + 62.83\,\text{Hz} \times 41.6\,\text{pF} \times 1 \times 10^{10}\,\Omega \times 0.001} = 426\,\text{ms} \tag{6.5.74}$$

The nominator in Equation 6.5.74 and input capacity C_V can be usually neglected as it generally applies that

$$\tau_V \approx R_G\, C_P \tag{6.5.75}$$

Now we can calculate the upper frequency limit of the responsivity:

$$f_V = \frac{1}{2\,\pi\,\tau_V} = \frac{1}{2\,\pi \times 0.426\,\text{s}} = 0.37\,\text{Hz} \tag{6.5.76}$$

For the specific noise equivalent power *NEP*, it applies according to Equation 5.2.3 that

$$NEP^* = \frac{\tilde{v}_{Rn}}{R_V} \tag{6.5.77}$$

Comparisons of pyroelectric sensors usually state specific detectivity D^* (Equation 5.3.2). For the current mode, it is

$$D_I^* = \frac{\alpha\,\pi_P\,T_R\,\sqrt{A_S^3}}{G_{th}} \frac{\omega_{Ch}}{\sqrt{1+\omega_{Ch}^2\tau_{th}^2}} \frac{1}{\sqrt{\tilde{i}_{Rn,P}^2 + \tilde{i}_{Rn,RV}^2 + \tilde{i}_{Rn,OV}^2 + \tilde{i}_{Rn,f}^2 + \dfrac{\tilde{v}_{Rn,OV}^2}{Z_f^2}}} \tag{6.5.78}$$

and for the voltage mode

$$D_V^* = \frac{\alpha\,\pi_P\,T_R\,\sqrt{A_S^3}}{G_{th}} \frac{\omega_{Ch}}{\sqrt{1+\omega_{Ch}^2\tau_{th}^2}} \frac{1}{\sqrt{\tilde{i}_{Rn,P}^2 + \tilde{i}_{Rn,RG}^2 + \tilde{i}_{Rn,OV}^2 + \dfrac{\tilde{v}_{Rn,OV}^2}{Z_V^2}}} \tag{6.5.79}$$

In order to assess the influence of individual noise sources, it is useful to state the individual components of the specific detectivity as figures of merit according to Equation 5.3.3.

In pyroelectric sensor elements, the contributions due to temperature fluctuation noise and the tanδ noise are of particular importance. Both components are independent of the electric connection of the sensor element. Regarding the contribution of the temperature fluctuation noise to specific detectivity, it applies that

$$D_T^* = \frac{\alpha\sqrt{A_S}}{\sqrt{4\,k_B\,T_S^2\,G_{th}}} \tag{6.5.80}$$

Inserting thermal conductance $G_{th,R}$ through radiation instead of thermal conductance G_{th}, we arrive at the BLIP detectivity (Equation 5.3.5).

The contribution of the specific detectivity through the tanδ noise results in

$$D_{\tan\delta}^* = \frac{\alpha\,\pi_P\,T_R\,\sqrt{A_S}}{G_{th}} \frac{\omega_{Ch}}{\sqrt{1+\omega_{Ch}^2\tau_{th}^2}} \frac{1}{\sqrt{4\,k_B\,T_S\,\omega_{Ch}\,C_P\,\tan\delta}} \tag{6.5.81}$$

Another important noise source in the input circuit is the noise of input resistance R_G (voltage mode) or R_V (current mode) of the series amplifier:

$$D_{VG}^* = \frac{\alpha \, \pi_P \, T_R \, \sqrt{A_S^3}}{G_{th}} \frac{\omega_{Ch}}{\sqrt{1 + \omega_{Ch}^2 \tau_{th}^2}} \sqrt{\frac{R_{V,G}}{4 \, k_B \, T_S}} \qquad (6.5.82)$$

with R_{VG} as resistance R_G (voltage mode) or R_V (current mode). Feedback resistance R_f also generates part of the specific detectivity:

$$D_f^* = \frac{\alpha \, \pi_P \, T_R \, \sqrt{A_S^3}}{G_{th}} \frac{\omega_{Ch}}{\sqrt{1 + \omega_{Ch}^2 \tau_{th}^2}} \sqrt{\frac{R_f}{4 \, k_B \, T_S}} \qquad (6.5.83)$$

The series amplifier contains two noise sources. The contribution to the specific detectivity due to voltage and current noise of the series amplifier

$$D_{VOV}^* = \frac{\alpha \, \pi_P \, T_R \, \sqrt{A_S^3}}{G_{th}} \frac{\omega_{Ch}}{\sqrt{1 + \omega_{Ch}^2 \tau_{th}^2}} \frac{Z_E}{\tilde{u}_{Rn,OV}} \qquad (6.5.84)$$

or

$$D_{IOV}^* = \frac{\alpha \, \pi_P \, T_R \, \sqrt{A_S^3}}{G_{th}} \frac{\omega_{Ch}}{\sqrt{1 + \omega_{Ch}^2 \tau_{th}^2}} \frac{1}{\tilde{i}_{Rn,OV}} \qquad (6.5.85)$$

Often the figures of merit are stated for certain frequency ranges. In the case of high chopper frequencies ($\omega_{Ch}^2 \gg 1/\tau_{th}^2$; responsivity according to Equation 6.5.66), Equations 5.5.81–6.5.85 become independent of chopper frequency f_{Ch}. Consequently, it results for the figures of merit that

$$D_{tan\delta}^* = \frac{\alpha \, \pi_P \, T_R}{c_S} \frac{1}{\sqrt{4 \, k_B \, T_S \, d_S \, \varepsilon \, \omega_{Ch} tan\delta}} \qquad (6.5.86)$$

$$D_{VG}^* = \frac{\alpha \, \pi_P \, T_R \sqrt{A_S}}{c_S \, d_S} \sqrt{\frac{R_{V,G}}{4 \, k_B \, T_S}} \qquad (6.5.87)$$

$$D_f^* = \frac{\alpha \, \pi_P \, T_R \sqrt{A_S}}{c_S \, d_S} \sqrt{\frac{R_f}{4 \, k_B \, T_S}} \qquad (6.5.88)$$

$$D_{VOV}^* = \frac{\alpha \, \pi_P T_R \sqrt{A_S}}{c_S d_S} \frac{Z_E}{\tilde{v}_{Rn,OV}} \qquad (6.5.89)$$

and

$$D_{IOV}^* = \frac{\alpha \, \pi_P \, T_R \sqrt{A_S}}{c_S \, d_S} \frac{1}{\tilde{i}_{Rn,OV}} \qquad (6.5.90)$$

Accordingly, it applies for the specific detectivity of the sensors in current mode that

$$D_{\mathrm{I}}^* = \frac{\alpha\,\pi_{\mathrm{P}} T_{\mathrm{R}} \sqrt{A_{\mathrm{S}}}}{c_{\mathrm{S}} d_{\mathrm{S}}} \frac{1}{\sqrt{\tilde{i}_{\mathrm{Rn,P}}^2 + \tilde{i}_{\mathrm{Rn,RV}}^2 + \tilde{i}_{\mathrm{Rn,OV}}^2 + \tilde{i}_{\mathrm{Rn,f}}^2 + \frac{\tilde{v}_{\mathrm{Rn,OV}}^2}{Z_{\mathrm{f}}^2}}} \tag{6.5.91}$$

and in voltage mode

$$D_{\mathrm{V}}^* = \frac{\alpha\,\pi_{\mathrm{P}} T_{\mathrm{R}} \sqrt{A_{\mathrm{S}}}}{c_{\mathrm{S}} d_{\mathrm{S}}} \frac{1}{\sqrt{\tilde{i}_{\mathrm{Rn,P}}^2 + \tilde{i}_{\mathrm{Rn,RG}}^2 + \tilde{i}_{\mathrm{Rn,OV}}^2 + \frac{\tilde{v}_{\mathrm{Rn,OV}}^2}{Z_{\mathrm{V}}^2}}} \tag{6.5.92}$$

Eventually, we can calculate *NETD* using Equation 5.4.6.

Example 6.15:   Specific Detectivity and *NETD* of a Pyroelectric Sensor

We will calculate specific detectivity D^* for the current and voltage mode of the sensors in Examples 6.5.1–6.5.5 (Figures 6.5.14 and 6.5.15). The BLIP detectivity is $1.81 \times 10^8 \, \mathrm{m\,Hz^{1/2}/W}$ and, in the example, corresponds to D_T^*.

Figure 6.5.16 shows temperature resolution *NETD* calculated from specific detectivity D^*. A comparison of the specific detectivity or *NETD*, respectively, between current and

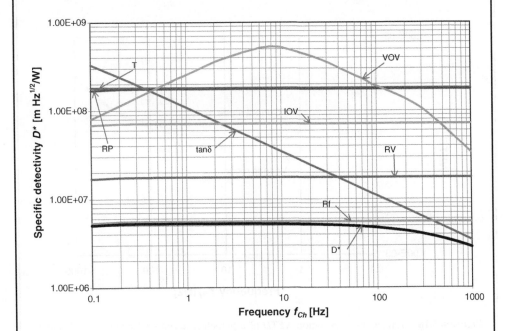

Figure 6.5.14   Specific detectivity D^* of a pyroelectric sensor in current mode; parameters: see Table 5.6.4

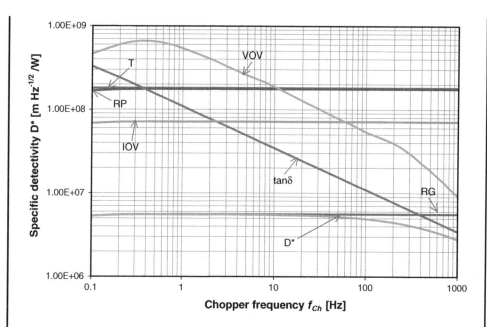

Figure 6.5.15 Specific detectivity D^* of a pyroelectric sensor in voltage mode; parameters: see Table 5.6.4

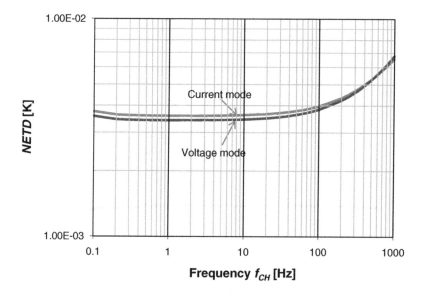

Figure 6.5.16 Temperature resolution *NETD* of a pyroelectric sensor. f-number $F = 1$; bandwidth $B = 100\,\text{Hz}$; wavelength range $\lambda = 8 \ldots 14\,\mu\text{m}$ ($I_M = 2.64\,\text{W}/(\text{m}^2\,\text{K})$)

voltage mode shows that both operating modes are actually equivalent. The differences lie in connection issues and in the frequency behaviour of the responsivity. In current mode, we can thus achieve large frequency-independent responsivity ranges, which is particularly useful for applications with chopper frequencies >100 Hz. Another advantage of the current mode is the small DC offset of the output signal due to contact issues.

6.5.3 Design of Pyroelectric Sensors

In addition to the pyroelectric detector element, a pyroelectric sensor almost always includes a series amplifier and the required maximum оHмic resistors (R_f or R_G). Only this way, we can implement a large signal-to-noise ratio. The structural designs of pyroelectric sensors are as numerous as their areas of application [16]. In the following, we will describe the structure of pyroelectric sensors as discrete components.

In voltage mode (Figure 6.5.17), usually a very low-noise field effect transistor (JFET) is integrated. In current mode (Figure 6.5.18), the housing includes a very low-noise operational amplifier together with a maximum оHмic resistor. There is no further external wiring connection required.

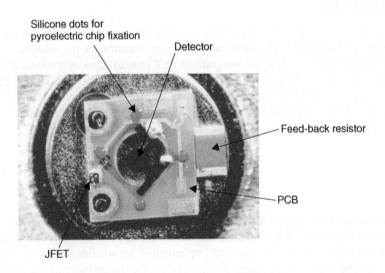

Figure 6.5.17 Pyroelectric sensor with intergrated junction field effect transistor (JFET). Chip area 3×3 mm^2; detector element area $\varnothing 1$ mm with black layer as absorber; $R_G = 100$ GΩ; housing TO39 (socket diameter 9.2 mm)

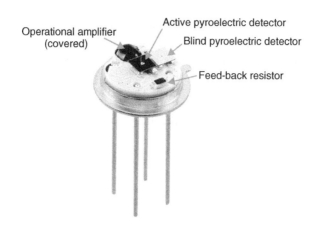

Figure 6.5.18 Temperature-compensated sensor in current mode. Typ LME-335; detector element area $A_S = 2 \times 2 \, \text{mm}^2$ with black layer; $R_f = 100 \, \text{G}\Omega$; $C_f = 0.2 \, \text{pF}$; housing TO39 (socket diameter 9.2 mm) (Reproduced by permission of InfraTec GmbH, Dresden, Germany)

Example 6.16: Responsivity of a Pyroelectric Sensor with Integrated FET

The responsivity of a pyroelectric sensor is calculated according to Equations 6.5.69 and 6.5.70. In Example 6.13, we assumed an amplification of the integrated operational amplifier with $a_v = 1$. If we integrate an FET instead of an operational amplifier, this assumption is no longer valid. The amplification depends then on the external wiring connection of the sensor. The simplest external wiring connection of the FET is a source follower circuit with resistance R_S (Figure 6.5.19a). For the voltage gain, it applies that

$$a_V \approx \frac{g_m R_S}{1 + g_m R_S} \tag{6.5.93}$$

with transconductance g_m of the FET. As source resistance R_S determines the output resistance of the circuit, it cannot adopt an arbitrarily large value. This means that the gain always is $a_V < 1$. Values that are implemented in practice lie in the range of 0.8–0.9. An important disadvantage of the circuit is the dependence of the gain on the operating point (drain current) and on the temperature. An advantage is the wiring connection of the FET with a current source (Figure 6.5.19b). In this case, the gain is $a_V \approx 1$.

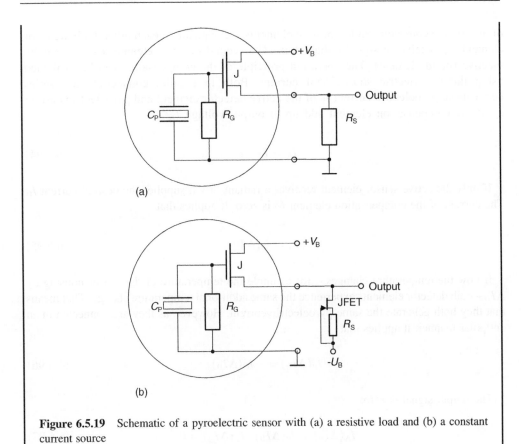

(a)

(b)

Figure 6.5.19 Schematic of a pyroelectric sensor with (a) a resistive load and (b) a constant current source

In order to compensate interferences of the ambient temperature, there is the option to calculate the difference in a series or parallel connection of two pyroelectric detector elements with only one element being exposed to infrared radiation (Figure 6.5.20). Here, the two detector elements are connected in antipolar manner, that is the directions

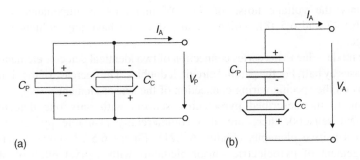

(a)   (b)

Figure 6.5.20 Compensated pyroelectric detectors in (a) parallel circuit and (b) series connection: C_P active pyroelectric detector element; C_C inactive pyroelectric detector element

of the spontaneous polarisation of both elements are opposite to each other. Only detector element C_P receives a signal radiant flux. The second detector element C_C is optically inactive (blind element). The operation principle of the compensation can be explained using the pyroelectric short-circuit current. Both in parallel connection and series connection, pyroelectric current I_P of the active detector element and pyroelectric current I_C of the compensation element add up to output voltage I_A:

$$I_A = I_P + I_C \tag{6.5.94}$$

If only the active sensor element receives a radiant flux it supplies the desired current I_P. The current of the compensation element I_C is zero. It applies that

$$I_A = I_P(\Delta\Phi) \tag{6.5.95}$$

If now the temperature changes – for example the temperature of the sensor housing by ΔT_0 – both detector elements experience the same additional temperature change. That means that they both generate the same pyroelectric current. However, as they are connected in an antipolar manner, it applies that

$$I_P(\Delta T_0) = -I_C(\Delta T_0) \tag{6.5.96}$$

The output signal is zero:

$$I_A(\Delta T_0) = I_P(\Delta T_0) - I_C(\Delta T_0) = 0 \tag{6.5.97}$$

or

$$I_A(\Delta\Phi, \Delta T_0) = I_P(\Delta\Phi) \tag{6.5.98}$$

In current mode, a parallel connection does not change responsivity. For calculating the noise we have to take into account the second pyroelectric detector element. However, only the voltage noise of the OV increases as impedance Z_P decreases (Equations 6.5.47 and 6.5.48), which actually does not have any influence on the total noise voltage.

In voltage mode – due to the series connection of two identical detector elements – element power decreases by half. For the parallel circuit, it doubles. The decrease of the signal-to-noise ratio depends on the specific wiring connection of the sensor element.

In addition to the described pyroelectric sensors with only one detector element (single-element sensors), there are also commercially available pyroelectric sensors with several sensitive elements (Figure 6.5.21). Figure 6.5.22 shows the image of a series arrangement of pyroelectric sensor elements with a pixel pitch of 50 μm. [21] presents a pyroelectric FPA (320 × 240 pixel) with a pixel pitch of 48.5 μm for thermographic applications.

Figure 6.5.21 Structures of pyrolelectric sensors. Left: single-channel; middle: four-channel; right: two-channel detectors (Reproduced by permission of InfraTec GmbH, Dresden, Germany)

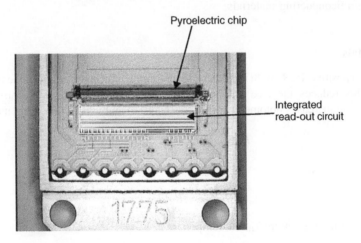

Figure 6.5.22 Pyroelectric LiTaO$_3$ line sensor. 128 sensor elements with $A_S = 90 \times 100\,\mu\text{m}^2$; pixel pitch $100\,\mu\text{m}$ (By courtesy of DIAS Infrared GmbH, Dresden, Germany)

6.6 Microbolometers

6.6.1 Principle

Bolometers are infrared sensors where temperature change ΔT_S of the sensor element due to a changed absorbed radiant flux $\Delta\Phi_S$ causes resistance change R_B. Figure 6.6.1 shows two basic circuits for determining the bolometer resistance. We evaluate voltage change ΔV_B caused by operating current I_B or ΔI_B flowing through bolometer resistance R_B.

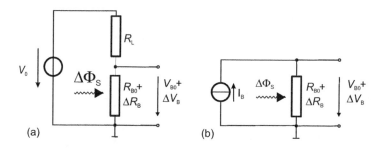

Figure 6.6.1   Basic circuits for bolometers: (a) with load resistance R_L and constant operating voltage V_0; (b) with constant current source I_B

The change of bolometer resistance ΔR_B caused by the temperature change is often described using temperature coefficient α_B (usually abbreviated as *TCR*: temperature coefficient of resistance):

$$\alpha_B = TCR = \frac{1}{R_B}\frac{dR_B}{dT_S} \tag{6.6.1}$$

The temperature dependence of the bolometer resistance is distinguished according to metals and semiconducting materials.

6.6.1.1 Metals

Metals have a positive TCR. With increasing temperature, lattice vibrations in the metal lattice decrease. This reduces the free pathlength of the electrons and consequently decreases conductivity and increases bolometer resistance. Close to ambient temperature, it applies that

$$R_B(T) = R_{B0}(1 + \alpha_B \Delta T_S) \tag{6.6.2}$$

with reference resistance

$$R_{B0} = R_B(T_0) \tag{6.6.3}$$

at operating point temperature T_0 and temperature change

$$\Delta T_S = T_S - T_0 \tag{6.6.4}$$

with T_S being the detector element temperature.

6.6.1.2 Semiconductors

The electric conductivity of semiconductors depends on the occupation of the conduction band. At absolute zero, the conduction band is empty; and the semiconductor is an insulator. With increasing temperature, electrons can move from the valence band to the conduction band. This means that there are freely movable electrons in the conduction band and holes in the valence

Table 6.6.1 *TCR* of different bolometer materials at ambient temperature

Metals		Semiconductors	
Material	α_B [%/K]	Material	α_B [%/K]
Ag	0.38	VO_x	−2.7
Al	0.39	Ni-Co-Mn-Oxid	−4.0
Au	0.34	YBaCuO	−3.5
Cu	0.39	GaAs	−9
Ni	0.60	a-Si[b]	−3.0
Ni-Fe[a]	0.23	a-Ge[b]	−2.1
Pt[a]	0.18	poly-Si: Ge	−1.4

[a]Thin layer.
[b]Amorphous semiconductor material.

band. Both contribute to conductivity. With increasing temperature, conductivity increases and bolometer resistance decreases resulting in a negative *TCR*. It applies in general that

$$R_B(T) = R_0\, e^{\frac{E_a}{k_B T}} \qquad (6.6.5)$$

with activation energy E_a, equal to half the bandgap energy, and reference resistance R_0. Using the definition of TCR in Equation 6.6.1, it results that

$$\alpha_B = \frac{1}{R_B}\frac{dR_B(T_S)}{dT_S} = -\frac{E_a}{k_B T_S^2} \qquad (6.6.6)$$

Table 6.6.1 states some *TCR* for different metals and semiconductor materials. Please note that thin layers typically have a lower *TCR* in comparison to volume materials. The *TCR* of metal thin layers is usually in the range of +0.2%/K. The *TCR* of semiconductors is heavily dependent on the manufacturing process, due to the different grain sizes, orientations and texture. The stated values are typical values.

6.6.2   Thermal Resolution

The resistance change caused by the temperature change is evaluated by measuring the voltage V_B (Figure 6.6.1b). For metal bolometers, the voltage change is

$$\Delta V_B = I_B R_{B0} \alpha_B \Delta T_S \qquad (6.6.7)$$

From there, we can calculate responsivity $R_{\Delta T}$ in relation to sensor temperature change ΔT_S:

$$R_{\Delta T} = \frac{d(\Delta V_B)}{d(\Delta T_S)} = I_B R_{B0} \alpha_B \qquad (6.6.8)$$

The voltage responsivity can be calculated using Equations 5.1.1 and 6.2.13:

$$R_V = \frac{R_{\Delta T}}{G_{th}\sqrt{1 + \omega_{Ch}^2 \tau_{th}^2}} = \frac{\alpha_B I_B R_{B0}}{G_{th}\sqrt{1 + \omega_{Ch}^2 \tau_{th}^2}} \qquad (6.6.9)$$

For semiconductor bolometers, it applies that

$$V_B = I_B \, R_0 \, e^{\frac{E_g}{k_B T_S}} = I_B \, R_B(T_S) \tag{6.6.10}$$

The responsivity is the slope of the V_B–T_S curve in Equation 6.6.10 and results with Equation 6.6.6 in

$$R_{\Delta T} = \frac{dV_B}{dT_S} = I_B \, R_B(T_S) \, \alpha_B \tag{6.6.11}$$

The voltage responsivity of semiconductor bolometers is

$$R_V = \frac{R_{\Delta T}}{G_{th}\sqrt{1 + \omega_{Ch}^2 \tau_{th}^2}} = \frac{\alpha_B I_B R_B}{G_{th}\sqrt{1 + \omega_{Ch}^2 \tau_{th}^2}} \tag{6.6.12}$$

In general, the voltage responsivity of a bolometer is – according to Equations 6.6.9 and 6.6.12 – directly proportional to bias current I_B.

Example 6.17:   Voltage Responsivity of a Bolometer

In the following examples, we will calculate the parameters of two microbolometers, with vanadium oxide (VO_x) and amorphous silicon (a-Si) being the resistor materials. These two types are currently the only commercially available microbolometer arrays. Example 6.21 shows the calculation of the thermal parameters (G_{th}, C_{th} and τ_{th}). With the characteristic values from Table 6.6.2, at $\omega_{Ch} = 0$ the voltage responsivity becomes

VO_x: $R_V = -328\,000$ V/W and
a-Si: $R_V = -732\,000$ V/W.

Example 6.17 shows that for negative *TCR* also the responsivity becomes negative. For calculating the temperature resolution, we only need the absolute value of the responsivity. Therefore, in the following, we will only use the absolute value of the responsivity.

When calculating the responsivity, we have to take into account that bolometer resistance R_B also is a function of bias current I_B and bolometer temperature T_S. In the static case, it follows directly from the thermal network in Figure 6.2.1 that

$$G_{th}(T_S - T_0) = i_B^2 R_B(T_S) + \Phi_S \tag{6.6.13}$$

Table 6.6.2   Characteristic values of the microbolometer examples

Material	VO_x		a-Si
TCR	−0.027/K		−0.03/K
G_{th}	2.05×10^{-7} W/K		1.02×10^{-7} W/K
R_B		50 kΩ	
I_B		50 μA	

Furthermore, OHM's law applies:

$$R_B = \frac{V_B}{I_B} \tag{6.6.14}$$

Inserting Equation 6.6.13 into Equation 6.6.4 and converting according to ΔT_S, we arrive at

$$\Delta T_S = T_S - T_0 = \frac{I_B V_B + \Phi_S}{G_{th}} \tag{6.6.15}$$

For metal bolometers, it results the current–voltage ratio

$$V_B(T_S) = R_{B0}\left(1 + \alpha_B \frac{I_B V_B + \Phi_S}{G_{th}}\right) \tag{6.6.16}$$

or resolving for V_B

$$V_B(I_B) = \frac{I_B R_{B0}\left(1 + \alpha_B \dfrac{\Phi_S}{G_{th}}\right)}{1 - \alpha_B I_B^2 \dfrac{R_{B0}}{G_{th}}} \tag{6.6.17}$$

For positive *TCR*, there results a pole:

$$1 - \alpha_B I_B^2 \frac{R_{B0}}{G_{th}} = 0 \tag{6.6.18}$$

In order for the measuring circuit not to become instable, bias current I_B of the bolometer must not exceed a certain value. Otherwise, the bolometer resistance would simply "blow!" In other words: If due to a temperature rise the resistance increases, the voltage – and consequently the power dissipation – increases too. Due to the positive *TCR*, this results in a further increase of the resistance and so on.

For a semiconductor bolometer, the following current–voltage relation applies

$$V_B = I_B R_{B0}\, e^{\frac{E_g}{2k_B\left(T_0 + \frac{I_B V_B + \Phi_S}{G_{th}}\right)}} \tag{6.6.19}$$

Unfortunately, this equation cannot be explicitly resolved for V_B or I_B. In the following example, we provide a numerical solution. For the calculation, we used Microsoft Excel Solvers.

Example 6.18: Current–Voltage Curve of Bolometers

Figure 6.6.2 presents the calculated *I/V* curve for metal and semiconductor bolometers. For metal bolometers, the pole point is located at a bias current of approximately $I_B = 70\,\mu A$. For semiconductor bolometers, the curve has its maximum at about $I_B = 6\,\mu A$.

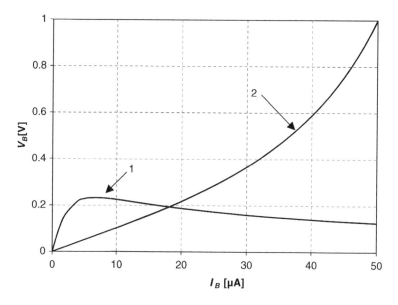

Figure 6.6.2 Current versus voltage for metal and semiconductor bolometers without IR irradiation ($\Phi_S = 0$) at ambient temperature ($T_S = 300$ K). The temperature conductance is $G_{th} = 10^{-7}$ W/K. Curve 1: semiconductor bolometer $\alpha_B = -0.03$/K; $R_B(300$ K$) = 100$ kΩ; Curve 2: metal bolometer $\alpha_B = 0.002$/K; $R_{B0} = 10$ kΩ

The dominating noise sources in a microbolometer are

- *1/f*-noise,
- resistance noise,
- temperature fluctuation noise.

They were presented in detail in Section 4.2. All named noise sources are located in the microbolometer bridge. They are statistically uncorrelated and their squared values can therefore be superimposed (addition of noise power). For low-noise read-out electronics, noise sources can be neglected for the subsequent signal processing.

Example 6.19:   Noise of a Microbolometer Pixel

We will calculate the noise of a microbolometer pixel for a semiconductor bolometer with vanadium oxide (VO$_X$) and amorphous silicon (a-Si) as resistor materials. The parameters in Table 6.6.3 will be applied to the microbolometer pixel.

For the following signal evaluation, the pixel signal will be integrated. The integration time is $t_I = 64$ μs, and applying Equation 4.1.85, the equivalent noise bandwidth becomes

$$B_{eq} = \frac{1}{2\,t_i} = 7.8 \text{ kHz} \qquad (6.6.20)$$

Table 6.6.3 Parameters applied in Example 6.19

Material	VO_x	a-Si
Pixel area A_P	$30 \times 30 \, \mu m^2$	
Emissivity ε	1	
Operating temperature T_S	300 K	
Bolometer resistance $R_B(300K)$	50 kΩ	
Load resistance R_L	50 kΩ	
Bias current i_B	10 μA	
Thermal conductance through heat conduction $G_{th,Leg}$ according to Example 6.21	2.05×10^{-7} W/K	1.02×10^{-7} W/K
Voltage responsivity R_V according to Example 6.17	328 000 V/K	732 000 V/K
Thermal time constant τ_{th}	3.5 ms	1.1 ms

We have presented the thermal noise of a microbolometer bridge already in Example 4.8. Neglecting the leakage current of the field effect transistor, the resistance noise becomes

$$\tilde{v}^2_{Rn.R} = 4\,k\,T(R_B \| R_L) \tag{6.6.21}$$

It applies for the $1/f$-noise of a microbolometer bridge according to Section 4.2.3 that

$$\tilde{v}^2_{Rn,1/f} = i_B^2\,R_B^2\,\frac{n}{f} \tag{6.6.22}$$

with material-dependent factor n. For

- vanadium oxide VO_x, it amounts to $n \approx 1 \times 10^{-13}$,
- amorphous silicon a-Si, it amounts to $n \approx 1 \times 10^{-11}$.

Integrating over the bandwidth $B = f_2 - f_1$, we arrive at the noise voltage we were looking for:

$$\tilde{v}_{R,1/f} = \sqrt{\int_{f_1}^{f_2} \tilde{v}^2_{Rn,1/f} df} = i_B\,R_B \sqrt{n \ln\frac{f_2}{f_1}} \tag{6.6.23}$$

The lower frequency limit f_1 is determined by observation time t_O:

$$f_1 \approx \frac{1}{4\,t_O} \tag{6.6.24}$$

For the upper frequency limit f_2, the equivalent noise bandwidth applies according to Equation 6.6.18:

$$f_2 = B_{eq} \tag{6.6.25}$$

The effect of observation or measuring time t_O on the flicker noise voltage is very low. In practice, the observation time is set as the time difference between two shutter processes, for example $t_B = 40$ min.

Table 6.6.4 Calculated noise of a microbolometer pixel

Bolometer material	a-Si	VO$_X$
Resistance noise $\tilde{v}_{R,R}$	1.80 µV	1.80 µV
1/f noise $\tilde{v}_{R,1/f}$	32.4 µV	3.2 µV
Temperature fluctuation noise $\tilde{v}_{R,T}$	7.87 µV	2.84 µV
Total noise $\tilde{v}_{R,R}$	33.3 µV	4.67 µV

The noise contribution of the temperature fluctuation noise is divided into the noise component of the radiation noise according to Equation 4.6.15 and the noise component due to heat conduction through the bridge legs according to Equation 4.6.12:

$$\overline{\Phi_{Rn,T}^2} = 16\,\varepsilon\,k_B\,\sigma\,T_S^5\,A_P + 4\,k_B\,G_{th,Leg}\,T_S^2 \tag{6.6.26}$$

As the noise bandwidth $B_{eq\Phi}$ of the temperature fluctuation noise is much smaller than electronic bandwidth B_{eq}, the noise bandwidth is determined by the thermal low-pass. According to Equation 4.6.7, it applies that

$$B_{eq\Phi} = \frac{1}{4\tau_{th}}$$

For VO$_x$, it applies that $B_{eq\Phi} = 71.4$ Hz. For a-Si, $B_{eq\Phi} = 227$ Hz, we apply the voltage responsivity and arrive at the noise voltage:

$$\tilde{v}_{Rn,T} = \overline{\Phi_{Rn,T}}\,R_V \tag{6.6.27}$$

The total noise is the square root of all noise components:

$$\tilde{v}_{R,B} = \sqrt{(\tilde{v}_{Rn,R}^2 + \tilde{v}_{Rn,1/f}^2)B_{eq} + \tilde{v}_{Rn,T}^2\,B_{eq\Phi}} \tag{6.6.28}$$

Table 6.6.4 summarizes all calculated values. For the presented microbolometers, the 1/f noise is the dominating noise source.

If we know the responsivity and noise values we can use it to calculate the noise-equivalent power *NEP*, detectivity D^* and temperature resolution *NETD*. The theoretical limits result from the corresponding BLIP values of the bolometer (see Chapter 5).

Example 6.20:   Temperature Resolution of a Microbolometer

We will apply Example 6.17 to calculate the temperature resolution for a microbolometer pixel. According to Equation 5.4.9, we assume for *NETD* an f-number of the optics of $F = 1$ and an operating wavelength range of 8–14 µm (value of differential exitance $I_M = 2.64$ W/(m^{2}K)). Table 6.6.5 states the calculated *NEP*, temperature responsivity R_T, specific detectivity D^* and *NETD*.

Table 6.6.5 Calculated temperature resolution *NETD* of a microbolometer at ambient temperature

Bolometer material	Equation	a-Si	VO$_X$
NEP	5.2.2	45.6 pW	14.2 pW
D*	5.3.2	5.8×10^7 m Hz$^{1/2}$/W	1.8×10^8 m Hz$^{1/2}$/W
R_T	5.1.35	348 µV/K	156 µV/K
NETD(300K)	5.4.9	96 mK	30 mK

6.6.3 Design of a Microbolometer Array

In order to achieve the required thermal isolation (see Section 6.2), the microbolometer resistors are manufactured as microbridges in vacuum. Figure 6.6.3 shows the simplified structure of such a pixel. The pixel areas typically lie in the range of between 15×15 µm^2 and 55×55 µm^2. Figure 6.6.4 is an image of the real bridge structure of an a-Si microbolometer. The four holes in the pixel are used for removing the sacrificial layer that is deposited during the manufacturing process of the bolometer layer.

The bridge structure consists of two support arms (legs) for the electrical connection, a membrane as the substrate layer und the bolometer resistance that is deposited on it. In addition, there are also other layers, for example for absorbing the incident IR radiation or for mechanically stabilizing the bridge (Figure 6.6.5). The thickness of the bridge amounts to a few 100 nm. In order to achieve the largest possible thermal isolation of the bolometer resistance, the entire structure is located in vacuum. The support arms contribute to thermal isolation, too. They connect the legs to the resistor area and serve as suspension of the bolometer resistance. The distance of the bridge to the chip surface or the reflector layer amounts to approximately 2.5 µm. Bridges and reflector (mirror on the read-out circuit) operate as optical resonators that absorb the IR radiation (λ/4-absorber) (see Example 2.2). The read-out electronics is situated under the bridge.

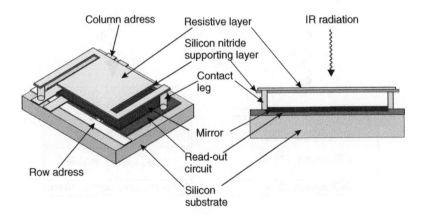

Figure 6.6.3 Structure of a microbolometer pixel (not to scale)

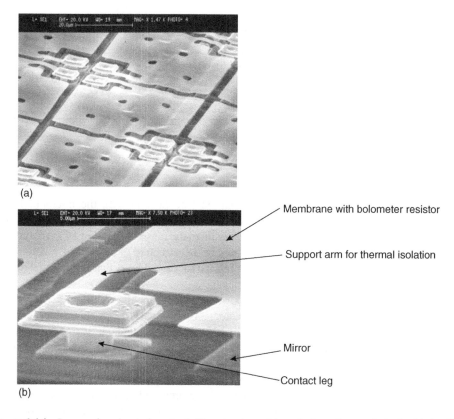

(a)

Membrane with bolometer resistor

Support arm for thermal isolation

Mirror

Contact leg

(b)

Figure 6.6.4 Image of a microbolometer bridge structure: (a) total view of several pixels, (b) detail of the image; pixel pitch 50 µm; bridge height 2.5 µm (By courtesy of ULIS, Veurey Voroize, France)

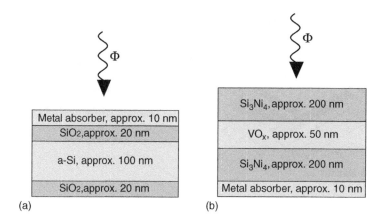

(a)   (b)

Figure 6.6.5 Layer systems of a microbolemeter bridge; (a) a-Si, (b) VO_x

Example 6.21:   Thermal Parameters of a Microbolometer Pixel

We will calculate the thermal parameters for the pixel geometry presented in Figure 6.6.6. In order to achieve minimum heat conduction through the support arms, these should be as long as possible. Taking into consideration half the bar area, the optical fill factor η amounts to approximately 0.64. The structural design should correspond to the one shown in Figure 6.6.5. For an emissivity ε of 0.8, the thermal conductance through radiation amounts according to Equation 6.1.14 to $G_{th,A} = 9 \times 10^{-9}$ W/K.

The thermal conductance of the support arms determines the heat conduction via the legs to the Si circuit as the feet already have to be considered thermal masses due to their large cross-section and their low height. Thermal conductance $G_{th,L}$ of the support arms is

$$G_{th,L} = 2\frac{\lambda_{th}A_S}{l_S} \tag{6.6.29}$$

with λ_{th} being the thermal conductivity, A_S the cross-section area and l_S the length of the plank. Factor 2 in Equation 6.6.29 takes into account that the pixel is mounted at two arms.

Table 6.6.6 shows some important material constants.

The heat conduction of the surrounding gas neglects the fact that the microbolomenter operates in the vacuum and the thermal capacity due to the residual gas is very low (Example 6.2). It applies to the total thermal conductance that

$$G_{th} = G_{th,A} + G_{th,S} \tag{6.6.30}$$

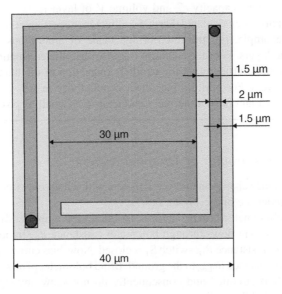

Figure 6.6.6   Geometry example of a microbolometer pixel. Pixel base area $40 \times 40\ \mu m^2$; pixel area $30 \times 30\ \mu m^2$; length of the support legs: $60\ \mu m$

Table 6.6.6 Material constants for thermal conductivity and the specific heat capacity of important materials for the design of bolometer pixels

Material	Thermal conductivity λ_{th} [W/(m K)]	Volumen-specific heat Capacity c'_p [Ws/(m^3 K)]
a-Si	30.0	1.6×10^6
SiO$_2$	0.9	2.6×10^6
Si$_3$N$_4$	3.2	1.6×10^6
VO$_X$	22.0	2.1×10^6

Table 6.6.7 Example results

Bolometer layer	a-Si	VO$_X$
G_{thA} in W/K	9×10^{-9}	9×10^{-9}
$G_{th,L}$ in W/K	2×10^{-7}	9.7×10^{-8}
G_{th} in W/K	2.1×10^{-7}	1.1×10^{-7}
C_{th} of the bolometer layer in Ws/K	1.4×10^{-10}	9.5×10^{-11}
C_{th} of the substrate layer in Ws/K	9.4×10^{-11}	2.9×10^{-10}
C_{th} in Ws/K	2.4×10^{-10}	3.9×10^{-10}
τ_{th} in ms	1.1	3.5

Heat capacity C_{th} of the pixel is the sum of the heat capacities of the individual layers:

$$C_{th} = \sum_i c'_{pi} V_i \qquad (6.6.31)$$

with volume-specific heat capacity c'_{pi} and volume V_i of layer i.

Table 6.6.7 compares the calculated thermal conductance coefficients and heat capacities of the described example. For the microbridge made of amorphous silicon, we have neglected the oxide layers. For the vanadium oxide bridge, the calculation of the thermal conductance takes into consideration the parallel connection of the thermal conductance of the individual layers. For both bridges, heat conductivity mainly occurs through the support arms. They should have a thermal conductance as small as possible.

6.6.4   Read-Out Electronics of Microbolometers

We measure the resistance change in microbolometers with a read-out circuit that is integrated in the silicon chip under the pixel (Figure 6.6.7).

Under each microbolometer bridge, there is an integrated amplifier J$_1$ and switch S$_1$. If switch S$_1$ is open no current flows through bolometer resistance R_B. In order to read out the pixel, that is bolometer resistance R_B, switch S$_1$ is closed. Now, bias current I_B flows via a blind bolometer pixel with resistance R_{blind} to the ground. Blind bolometer pixels are 'normal' pixels, that are thermally short-circuited and consequently do not show any signal change. They compensate for unwanted fluctuations of the operating temperature and balance technology-based tolerances of the resistors. Bolometer and blind bolometer resistance R_B or R_{blind} form

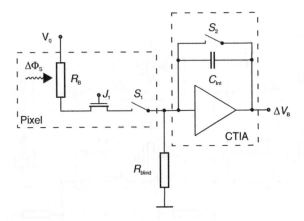

Figure 6.6.7   Electronics for the evaluation of resistance change in bolometer pixels [17]

a half-bridge circuit. If due to a resistance change, current I_B changes, this change ΔI_B is integrated by a capacitive transimpedance amplifier (CTIA). After integration time t_{Int} has elapsed, signal voltage ΔV_B is available as output signal of the CTIA. The integration time depends on the refresh rate and the number of pixels. In the above example, the integration time was set at $t_{Int} = 64\ \mu s$. After completed integration, the output signal is sampled and the CTIA is reset via switch S_2. Then the next pixel is read out.

Figure 6.6.8 presents the concept of an $m \times n$ microbolometer array that has m columns in n rows.

Table 6.6.8 states common image formats and the resulting pixel frequencies and integration times. As the read-out circuit needs for its internal organisation additional clocks between the rows and frames, the frequencies constitute minimum values. In Europe, the common image frequency is $f_B = 50\ Hz$, in the US, it is $f_B = 60\ Hz$. This corresponds to the respective television standards. Also the array formats follow television standards. The maximum integration time is calculated applying frame time t_F and row number n

$$t_{Int} < \frac{t_F}{n} \tag{6.6.32}$$

The pixels of a row are read-out simultaneously. This requires m integrators (CTIA). After completing the integration, the output signals are buffered in a sample-and-hold amplifier (S&H) and thus are available to the multiplexer during the integration of the next column. The multiplexer then provides the m pixels of the previous row serially at the output. After n read-out cycles, the image is completely read out.

Due to the read-out procedure, the temperature distribution in a frame is not simultaneously recorded (Figure 6.6.9). Signal recording, that is the integration of the first row, starts at time t_0 and is completed at time $t_0 + t_{Int}$. At this time, reading out of the first row and integration of the second row and so on starts. In the image recorded in a microbolometer, the first row contains the temperature information during time interval t_0 to $t_0 + t_{Int}$. As opposed to this, the last row contains the temperature information relating to time interval from $t_0 + (n-1)t_{Int}$ to $t_0 + nt_{Int}$. The time difference actually corresponds to frame time $t_F = 1/f_F$ of 20 or 16.7 ms. This read-out procedure is called rolling frame.

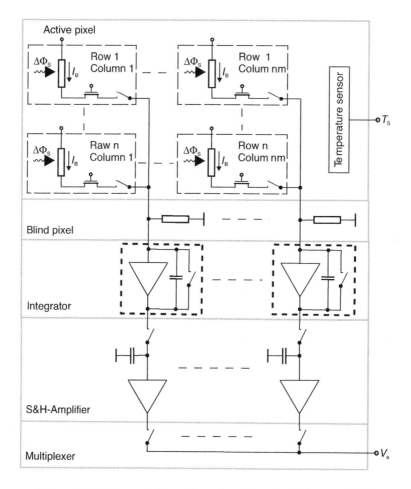

Figure 6.6.8 Concept of a read-out circuit for microbolometers [17]

Table 6.6.8 Image formats and resulting minimum pixel frequencies of microbolometer arrays (m number of columns; n number of rows)

Format $m \times n$	Number of pixels	Minimum pixel frequency		Maximum integration time	
		$f_B = 50$ Hz	$f_B = 60$ Hz	$f_B = 50$ Hz	$f_B = 60$ Hz
160×120	19 200	0.9600 MHz	1.15 200 MHz	167 μs	139 μs
320×240	76 800	3.8400 MHz	4.60 800 MHz	83 μs	69 μs
384×288	110 592	5.5296 MHz	6.63 552 MHz	69 μs	58 μs
640×480	307 200	15.3600 MHz	18.43 200 MHz	42 μs	35 μs
1024×768	786 432	39.3216 MHz	47.18 592 MHz	26 μs	22 μs
1920×1080	2 073 600	103.68 MHz	124.41 MHz	18.5 μs	15.4 μs

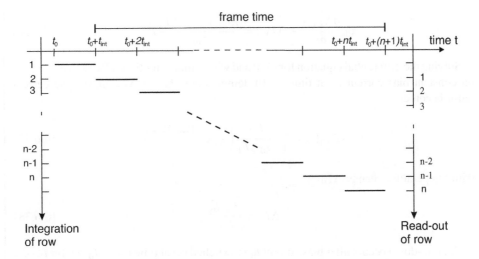

Figure 6.6.9   Integration and read-out time of the rows in an $m \times n$ FPA of a bolometer array

---

Example 6.22:   Temperature Change of a Bolometer Due to the Bias Current

Bias current I_B of a pixel is only connected during the integration for read-out purposes (pulse mode). This has the advantage that the power consumption of the entire array is low. However, during the current pulse, there occurs a rather considerable electric power draw that translates into an undesired temperature change in the pixel.

We use the simple thermal model in Section 6.2 to calculate this temperature change (Figure 6.2.1). Taking into consideration electric power P_E, power P_S fed into the sensor becomes

$$P_S = \alpha\,\Delta\Phi_S(t) + P_E(t) \tag{6.6.33}$$

For the electric power, it applies that

$$P_E(t) = I_B^2(t)R_B(t) = R_{B0}(1 + \alpha_B \Delta T)\,I_B^2(t) \tag{6.6.34}$$

Thus it follows from Equation 6.2.3:

$$\alpha\,\Delta\Phi_S(t) + P_E(t) = C_{th}\frac{d[\Delta T_S(t)]}{dt} + G_{th}\,\Delta T_S(t) \tag{6.6.35}$$

As the electric power and the absorbed radiation power are independent of each other, the above differential equation can be resolved separately for both excitations and the resulting temperature differences can be superimposed. In order to calculate the temperature change due to the bias current, in the following the absorbed radiation power will be set to zero ($\alpha\,\Delta\Phi_S(t) = 0$). Equation 6.6.35 thus becomes

$$\tau_{\text{th}} \frac{\mathrm{d}\Delta T(t)}{\mathrm{d}t} + \Delta T(t) = \frac{R_{\text{B0}}(1 + \alpha_{\text{B}}\Delta T)}{G_{\text{th}}} I_{\text{B}}^2(t) \qquad (6.6.36)$$

Solving the differential equation for $t \geq 0$ and with initial condition $\Delta T(0) = 0$ (switching on constant bias current I_{B} at time $t = 0$), temperature change ΔT_{on} during the current pulse becomes

$$\Delta T_{\text{on}}(t) = \frac{\Delta T_0}{1 - \alpha_{\text{B}}\Delta T_0} \left(1 - e^{-\frac{1 - \alpha_{\text{B}}\Delta T_0}{\tau_{\text{th}}}t}\right) \qquad (6.6.37)$$

with temperature change ΔT_0

$$\Delta T_0 = \frac{R_{\text{B0}} I_{\text{B}}^2}{G_{\text{th}}} \qquad (6.6.38)$$

Cooling-down occurs after bias current I_{B} is switched off at time $t = t_0$ ($I_{\text{B}} = 0$ for $t > t_0$). For temperature change ΔT_{off} after switching off the current pulse, it applies that

$$\Delta T_{\text{off}}(t) = \Delta T_{\text{on}}(t_0) \, e^{-\frac{t - t_0}{\tau_{\text{th}}}} \qquad (6.6.39)$$

Figure 6.6.10 presents the time curve of the pixel temperature change. Due to the temperature-dependent resistance, the thermal time constant is reduced by factor $1 - \alpha_{\text{B}}\Delta T_0$ during the heating of the pixel. This means that it warms up faster than it cools down. The effect that the temperature change due to the bias current has on the sensor parameters must be included into the calculation. Kruse and Skatrud [18] show corresponding example calculations.

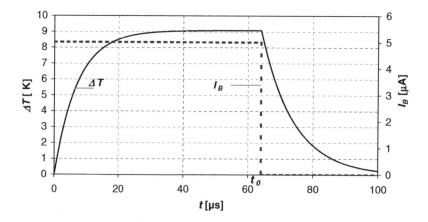

Figure 6.6.10 Pixel temperature curve for a pulse-formed bias current I_{B}. Parameters: $\tau_{\text{th}} = 10 \, \text{ms}$; $I_{\text{B}}(0 \ldots 64 \, \mu s) = 5 \, \mu A$; $R_B = 50 \, k\Omega$; $G_{\text{th}} = 10 \times 10^{-7} \, \text{W/K}$; $\alpha_{\text{B}} = -0.03/\text{K}$; $\Delta T_0 = 12.5 \, \text{K}$

In addition to the microbolometer array, a complete microbolometer array sensor can include further components, such as a temperature sensor, a PELTIER element (thermoelectric cooler TEC) and a vacuum getter.

The temperature sensor is used to determine the bolometer temperature and is integrated into the read-out circuit (Figure 6.6.8). In connection with a TEC, the sensor temperature can be set at a constant value. Here the admissible deviation from the rule is smaller than 0.1 K. If we select a sensor temperature in the ambient temperature range of between 20 and 30 °C we can achieve low power consumption. For low-cost sensors no TEC is used. These sensors are called TEC-less microbolometers.

To maintain the vacuum, the housing includes so-called getters made of reactive materials that absorb the residual gases. In order to activate the getters we usually have to heat them with the electric current. For that purpose, there are special pins at the housing. The getters rarely have to be activated.

Figure 6.6.11 shows the chip image of a microbolometer array and a complete housing. The sensor windows are exclusively made of germanium, as germanium has optimum optical properties in the operating wavelength range of about 8–14 μm. It usually is 1 mm thick. The transmission of the sensor window depends on the anti-reflection coating of the germanium. Usually, there are two different kinds of anti-reflection coating:

- broadband anti-reflection coating,
- bandpass anti-reflection coating.

Figure 6.6.12 presents two typical transmission curves. The bandpass anti-reflection coating has the advantage that it does not require any further filters in the optical channel. For manufacturers, the broadband anti-reflection coatings have the advantage that the sensor has a considerably better thermal resolution (reduction of the *NETD* value to about 50%). It is essential for temperature-measuring camera systems that radiation below a wavelength of 7.5 μm is blocked, as in the wavelength range of between 5 to 7.5 μm, there is a strong absorption of infrared radiation by humidity in the atmosphere.

(a)   (b)

Figure 6.6.11 (a) Chip image of a microbolometer array and (b) microbolometer array sensor (By courtesy of ULIS, Veurey Voroize, France)

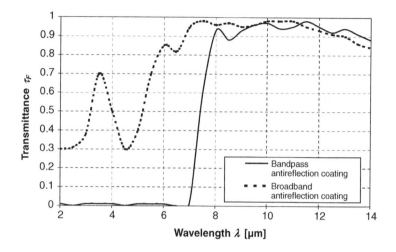

Figure 6.6.12   Typical window transmission curves [19, 20]

6.7   Other Thermal Infrared Sensors

6.7.1   Bimorphous Infrared Sensors

Due to their different coefficients of thermal expansion, bimetals or bimorphs are deformed during a temperature change, that is they are expanded or bent. This effect can be used in micromechanical bimorph elements for measuring the temperature change due to an incident IR radiation. Such sensors are called microcantilever detectors. The deflection can be measured capacitively [22], piezoresistively [23] or optically [31].

For a capacitive signal read-out, such as presented in Figure 6.7.1, sensor capacity ΔC_S is changed when the upper electrode is lifted by Δz and tilted by angle $\Delta \varphi$.

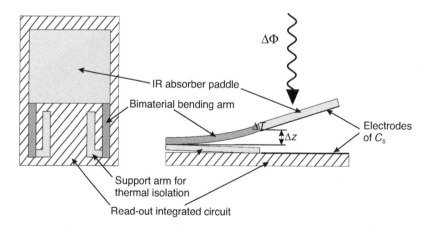

Figure 6.7.1   Micromechanical cantilever temperature sensor with capacitive read-out

Example 6.23:   Sensor Capacity for a Tilted Electrode

We will calculate the capacity of the electrode arrangement according to Figure 6.7.1 in relation to angle φ. In general, it applies for the capacity that

$$C_S = \varepsilon_0 \int_{A_S} \frac{dx\,dy}{z(y)} \tag{6.7.1}$$

The electrodes have the area

$$A_S = ab \tag{6.7.2}$$

By tilting the upper electrode, it applies for electrode distance z that

$$z(y) = z_0 + y \tan \varphi \tag{6.7.3}$$

with z_0 as the electrode distance without deflection. It thus results from Equation 6.7.1 that

$$C_S = \varepsilon_0 \frac{b}{z_0} \int_0^a \frac{dy}{1 + \dfrac{\tan \varphi}{z_0} y} \tag{6.7.4}$$

We assume small angles φ in order to be able to neglect the reduction of the effective capacitor area under deflection. The solution of the integral is

$$C_S = \varepsilon_0 \frac{b}{\tan \varphi} \ln\left(1 + \frac{a}{z_0} \tan \varphi\right) \tag{6.7.5}$$

It applies for small angles $\varphi \ll 1$ that

$$\tan \varphi \approx \varphi \tag{6.7.6}$$

and

$$\ln(1 + x) \approx x \tag{6.7.7}$$

with $x = \dfrac{a}{z_0} \tan \varphi \ll 1$. Then, Equation 6.7.5 results in the familiar formula for plate capacitors:

$$C_S = \varepsilon_0 \frac{ab}{z_0} \tag{6.7.8}$$

It applies for deflection Δz under the conditions that both bimaterial layers have the same thickness s and that both materials have the same modulus of elasticity [1]:

$$\Delta z = \frac{3\,l^2 \Delta\alpha_M \Delta T}{8\,s} \tag{6.7.9}$$

with the length of the bimaterial levers l and difference $\Delta\alpha_M$ of thermal expansion coefficients α_{M1} and α_{M2} of the bimaterials:

$$\Delta\alpha_M = \alpha_{M1} - \alpha_{M2} \qquad (6.7.10)$$

For small temperature changes we can neglect the tilting of the electrode; and Equation 6.7.8 can be applied to the capacity. For the capacity change around operating point z_0, it then applies that

$$\frac{dC_S}{dz} = -\frac{\varepsilon_0 A}{z_0^2} = -\frac{C_{S0}}{z_0} \qquad (6.7.11)$$

with sensor capacity C_{S0} at the operating point. It follows from Equation 6.7.9 that

$$dz = \frac{3 l^2 \Delta\alpha_M}{8\,s} dT \qquad (6.7.12)$$

Thus, Equation 6.7.11 becomes

$$\frac{dC_S}{dT} = -\frac{C_{S0}}{z_0}\frac{3\,l^2\Delta\alpha_M}{8\,s} \qquad (6.7.13)$$

In equivalence to *TCR* of resistive sensors (bolometers, see Section 6.6), we can define a temperature coefficient of capacitance (*TCC*) to describe capacitive infrared sensors [24]:

$$TCC = \frac{1}{C_S}\frac{dC_S}{dT} \qquad (6.7.14)$$

TCC of the above calculated arrangement is

$$TCC = -\frac{3}{8}\frac{l^2\Delta\alpha_M}{s\,z_0} \qquad (6.7.15)$$

That means that temperature coefficient *TCC* depends on difference $\Delta\alpha_M$ of the thermal expansion coefficients of the bimaterials and on the geometrical structure.

The half-bridge evaluation circuit in Figure 6.7.2 is used to calculate the voltage responsivity. Output voltage change ΔV_S is

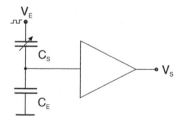

Figure 6.7.2   Circuit for measuring capacity change

$$\Delta V_S = V_E \frac{\Delta C_S}{C_E} \tag{6.7.16}$$

with pulse-type bias voltage V_E and capacity C_E, that is composed of the bridge capacity and the input capacity of the subsequent amplifier circuit – usually an integrator. Voltage V_E cannot be set at any arbitrary value as electrostatic forces attract the electrodes to each other.

Capacity change due to temperature change can be calculated using Equation 6.7.11 with the transition to the difference quotient:

$$\Delta C_S = -C_{S0} \frac{\Delta z}{z_0} \tag{6.7.17}$$

Thus the voltage change becomes

$$\Delta V_S = -V_E \frac{C_{S0}}{C_E} \frac{\Delta z}{z_0} = -V_E \frac{C_{S0}}{C_E} \frac{3 \, l^2 \Delta \alpha_M}{8 \, s \, z_0} \Delta T \tag{6.7.18}$$

Using Equation 6.2.4, we can now calculate the voltage responsivity:

$$R_V = \frac{\alpha}{G_{th}} \frac{V_E C_{S0}}{C_E} \frac{3 l^2 \alpha_M}{8 \, s \, z_0} \tag{6.7.19}$$

The left-hand term of the responsivity describes the absorption of the radiation and the thermal isolation. The middle term contains the electronic signal evaluation whereas the right-hand term describes temperature coefficient TCC according to Equation 6.7.15.

The dominating electric noise source is the kTC noise (Equation 4.2.10):

$$\tilde{v}_{R,a}^2 = \frac{k_B T_S}{C_{S0} + C_E} \tag{6.7.20}$$

Neglecting other noise sources, it applies to NEP that

$$NEP = \sqrt{\frac{k_B T_S}{C_{S0} + C_E} \frac{G_{th}}{\alpha} \frac{C_E}{V_E C_{S0}} \frac{8 \, s \, z_0}{3 l^2 \alpha_M}} \tag{6.7.21}$$

Example 6.24:   Temperature Resolution of a Cantilever

A pixel with an area of $50 \times 50 \, \mu m^2$ is suspended by two cantilevers of a length of $100 \, \mu m$ and a width of $2 \, \mu m$ each. The cantilevers are bent (Figure 6.7.1), where half a cantilever each is used for isolation purposes and the other half consists of bimaterial. In the operating point, we find pixel $z_0 = 2.5 \, \mu m$ over the read-out circuit. With Equation 6.7.8, capacity C_{S0} becomes

$$C_{S0} = \frac{\varepsilon_0 (50 \, \mu m)^2}{2.5 \, \mu m} = 8.85 \times 10^{-15} \, F \tag{6.7.22}$$

We assume capacity C_E to be identical:

$$C_E = C_{S0} \tag{6.7.23}$$

The bimaterial cantilevers consist of aluminium (Al; coefficient of thermal expansion $\alpha_{M1} = 25$ ppm/K) and silicon oxide (SiO$_2$; $\alpha_{M2} = 0.35$ ppm/K). Both materials are 0.2 µm thick. The resulting temperature coefficient is

$$TCC = -\frac{3}{8}\frac{(50\,\mu\text{m})^2}{0.2\,\mu\text{m} \times 2.5\,\mu\text{m}}(25 \times 10^{-6} - 0.35 \times 10^{-6})/\text{K} = -4.62\%/\text{K} \tag{6.7.24}$$

Thermal conductance G_{th} consists of the two bimaterial cantilevers and the two support arms:

$$\frac{1}{G_{th}} = \frac{1}{2\,G_{th,Al}} + \frac{1}{2\,G_{th,iso}} \tag{6.7.25}$$

In the bimaterial cantilever, the aluminium layer determines the thermal conductance

$$G_{th,Al} = \frac{\lambda_{th,Al}A_{Arm}}{l_{Arm}} = \frac{237\,\frac{\text{W}}{\text{mK}}(0.2\,\mu\text{m} \times 2\,\mu\text{m})}{50\,\mu\text{m}} = 1.9 \times 10^{-6}\,\frac{\text{W}}{\text{K}} \tag{6.7.26}$$

For identical geometric dimensions, it applies to the insulation arm that

$$G_{th,iso} = \frac{\lambda_{th,SiO}A_{iso}}{l_{iso}} = \frac{0.9\,\frac{\text{W}}{\text{mK}}(0.2\,\mu\text{m} \times 2\,\mu\text{m})}{50\,\mu\text{m}} = 7.2 \times 10^{-9}\,\frac{\text{W}}{\text{K}} \tag{6.7.27}$$

The entire temperature difference actually occurs over the isolation arms. The above assumption that the bimaterial cantilevers warms evenly coincides with the real thermal conditions. It applies that:

$$G_{th} = 2\,G_{th.iso} \tag{6.7.28}$$

With absorption coefficient $\alpha = 1$ and operating voltage $V_E = 1$V, NEP then becomes

$$NEP = \sqrt{\frac{k_B T_S}{2\,C_{S0}}\frac{G_{th}}{U_E}\frac{8\,s\,z_0}{3l^2\alpha_M}}$$

$$= \sqrt{\frac{k_B \times 300\,\text{K}}{28.85 \times 10^{-15}\,\text{F}}\frac{2 \times 7.2 \times 10^{-9}\frac{\text{W}}{\text{K}}}{1\,\text{V}}\frac{8 \times 0.2\,\mu\text{m} \times 2.5\mu\text{m}}{3 \times (50\,\mu\text{m})^2(25 \times 10^{-6} - 0.35 \times 10^{-6})/\text{K}}}$$

$$= 1.16 \times 10^{-10}\,\text{W}$$

$$\tag{6.7.29}$$

NETD of the cantilever sensor thus is calculated according to Equation 5.4.9 applying *NEP* and an f-number of $F = 1$ in the wavelength range between 8 and 14 μm at 27 °C; and it becomes

$$NETD = \frac{4F^2 + 1}{A_S} \frac{NEP}{I_M} = \frac{5}{(50\,\mu\text{m} \times 50\,\mu\text{m})} \frac{1.16 \times 10^{-8}\text{W}}{2.64\,\text{W}/(\text{m}^2\text{K})} = 0.09\,\text{K} \qquad (6.7.30)$$

The deflection of the cantilever at noise-equivalent temperature difference *NETD* is according to Equation 6.7.9

$$\Delta z = \frac{3 \times (50\,\mu\text{m})^2 (25 \times 10^{-6} - 0.35 \times 10^{-6})/\text{K}}{8 \times 0.2\,\mu\text{m}} 0.09\,\text{K} = 10.5\,\text{nm} \qquad (6.7.31)$$

Finally, we want to estimate the thermal time constant. For this purpose, we will calculate the heat capacity. It is composed of the heat capacities of both bimaterial cantilevers and that of the absorber area:

$$C_{th} = 2C_{th,bi} + C_{th,Ab} \qquad (6.7.32)$$

The heat capacity of the bimaterial cantilevers is calculated using volume-specific heat capacities c' and the volumes V of both materials Al and SiO$_2$:

$$C_{th,bi} = c'_{Alu} V_{Alu} + c'_{SiO} V_{SiO} \qquad (6.7.33)$$

In the example, the two volumes are identical:

$$V_{Alu} = V_{SiO} = 50\,\mu\text{m} \times 2\,\mu\text{m} \times 0.2\,\mu\text{m} = 2 \times 10^{-17}\,\text{m}^3 \qquad (6.7.34)$$

Thus, the heat capacity of the bimaterial cantilevers becomes

$$C_{th,bi} = 2 \times 10^{-17}\,\text{m}^3 \left(2.4 \times 10^6 \frac{\text{Ws}}{\text{K} \times \text{m}^3} + 2.6 \times 10^6 \frac{\text{Ws}}{\text{K} \times \text{m}^3} \right) = 4.86 \times 10^{-11} \frac{\text{Ws}}{\text{K}}$$
$$(6.7.35)$$

The absorber areas basically consist of silicon nitride (Si$_3$N$_4$) with a thickness of 0.2 μm. We can neglect the thin metal layers that are applied for absorption purposes. It applies that

$$C_{th,Ab} = c'_{Si_3N_4} V_{Ab} = 1.6 \times 10^6 \frac{\text{Ws}}{\text{K} \times \text{m}^3} (50\,\mu\text{m})^2 \times 0.2\,\mu\text{m} = 8 \times 10^{-10} \frac{\text{Ws}}{\text{K}} \qquad (6.7.36)$$

That means that the heat capacity is determined by the absorber area. The time constant then becomes

$$\tau_{th} = \frac{C_{th}}{G_{th}} = \frac{8 \times 10^{-10} \frac{\text{Ws}}{\text{K}}}{1.44 \times 10^{-8} \frac{\text{W}}{\text{K}}} = 55\,\text{ms} \qquad (6.7.37)$$

6.7.2   Micro-GOLAY Cells

A GOLAY cell consists of a closed, thermally well-isolated gas volume (cell) that increases its temperature due to the absorption of radiation [25]. Usually, we measure the deflection of the cell wall that is deformed due the increased pressure and consequently we can calculate the temperature change of the gas. The GOLAY cell is also an opto-acoustic or optopneumatic sensor. In classical GOLAY cells, the deflection of the cell wall is determined with optical procedures, for example laser interferometry. Therefore, these sensors are large and not very robust.

Miniaturised GOLAY cells are manufactured using micromechanical procedures. Figure 6.7.3 shows such a version. In a base body (silicon rim), there is an opening that is closed above with a stiff cover and below with a flexible membrane (bending plate). The upper membrane, for example made of Si_3N_4/SiO_2, absorbs together with an absorber (metallisation, e.g. Pt) the IR radiation. The bending plate, for example made of Si_3N_4, is deformed by the changed pressure of the confined gas. The substrate and lower side of the bending plate are metallised. Their distance amounts to only a few micrometers. Bending plate and substrate form changeable capacitor C_G. The deformation of the bending plate results in capacity change ΔC_G [26].

There are also other methods for determining the deflection of the bending plate, for example measuring the tunnel current between bending plate and substrate [27] or evaluating the deflection using piezoresistors.

The law of the ideal gas states that

$$pV = nRT \tag{6.7.38}$$

with pressure p in volume V, general gas constant R and amount of substance n. We are now looking at small changes around operating point p_0, V_0 and T_0. For this purpose, we derive Equation 6.7.38 with respect to T:

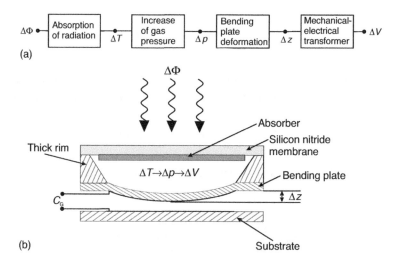

Figure 6.7.3   (a) Principle and (b) structure of a miniaturised GOLAY cell with capacitive read-out

$$\frac{dp}{dT}V + \frac{dV}{dT}p = nR \tag{6.7.39}$$

By transformation, we arrive at

$$\frac{dp}{p} + \frac{dV}{V} = \frac{nR}{pV}dT \tag{6.7.40}$$

Inserting Equation 6.7.38, it results that

$$\frac{dp}{p} + \frac{dV}{V} = \frac{dT}{T} \tag{6.7.41}$$

With the transition to the difference quotients, it becomes

$$\frac{\Delta p}{p_0} + \frac{\Delta V}{V_0} = \frac{\Delta T}{T_0} \tag{6.7.42}$$

Absorbed radiant flux $\Delta\Phi$ determines the temperature change. The change in volume due to increased pressure can be calculated using the deformation of the bending plate. For deformation $w(x,y)$ of square plates with an edge length of $2a$, it approximately applies that [28]

$$w(x,y) = \frac{0.383\,\Delta p}{18\,a^4 D}(x^2-a^2)^2(y^2-a^2)^2 \tag{6.7.43}$$

with bending strength D:

$$D = \frac{Ed^3}{12\,(1-v)} \tag{6.7.44}$$

as well as membrane thickness d, YOUNG's modulus E and POISSON's ratio v. With deformation $w(x,y)$, it results for the volume change that

$$\Delta V = \int_A w(x,y)\,dx\,dy \tag{6.7.45}$$

As the pressure is independent of the position, the integral in Equation 6.7.45 can be easily solved:

$$\Delta V = \frac{0.383\,\Delta p}{18\,a^4 D} \int_{-a}^{+a}\int_{-a}^{+a} (x^2-a^2)^2(y^2-a^2)^2 dx\,dy \tag{6.7.46}$$

It applies for the integral over dx:

$$\int_{-a}^{+a} (x^2-a^2)^2 dx = \frac{1}{5}x^5 - \frac{2}{3}a^2 x^3 + a^4 x \Big|_{-a}^{a} = \frac{16}{15}a^5 \tag{6.7.47}$$

For the integral over dy, we arrive at the same result. Thus, it applies to the volume change that

$$\Delta V = \frac{0.383\,\Delta p}{18\,a^4 D}\frac{256}{225}a^{10} = 2.42 \times 10^{-2}\frac{a^6}{D}\Delta p \qquad (6.7.48)$$

Now we use Equations 6.7.42 and 6.7.48 to calculate pressure change Δp as a function of temperature change ΔT:

$$\Delta p = \frac{p_0}{T_0}\frac{\Delta T}{1 + \dfrac{p_0}{V_0}\left(2.42 \times 10^{-2}\dfrac{a^6}{D}\right)} \qquad (6.7.49)$$

It follows from Equations 6.7.48 and 6.7.49 that the volume change is proportional to the temperature change:

$$\Delta V = W\,\Delta T \qquad (6.7.50)$$

with the temperature-related volume change

$$W = 2.42 \times 10^{-2}\frac{a^6}{D}\frac{p_0}{T_0}\frac{1}{1 + \dfrac{p_0}{V_0}\left(2.42 \times 10^{-2}\dfrac{a^6}{D}\right)} = \frac{p_0}{T_0}\frac{1}{41.3\dfrac{D}{a^6} + \dfrac{p_0}{V_0}} \qquad (6.7.51)$$

For capacity C_G, it generally applies that

$$C_G = \varepsilon_0 \int_A \frac{dxdy}{z_0 - w(x,y)} \qquad (6.7.52)$$

Using deformation $w(x,y)$ and Equation 6.7.14 we can calculate temperature coefficient TCC of capacity C_G. For capacity C_{G0} around the operating point ($\Delta T = 0$ and thus even $w(x,y) = 0$), it applies that

$$C_{G0} = \varepsilon_0 \frac{a^2}{z_0} \qquad (6.7.53)$$

Area-related capacity C_G'' is

$$C_G'' = \frac{\varepsilon_0}{z_0} \qquad (6.7.54)$$

It follows for the capacity change that

$$\frac{dC_G''}{dz} = -\frac{\varepsilon_0}{z_0^2} \qquad (6.7.55)$$

Change dz corresponds to deformation $w(x,y)$ in Equation 6.7.43:

$$dz = w(x,y) \qquad (6.7.56)$$

We can use Equation 6.7.55 to integrate over capacitor area A_G and thus calculate the capacity change:

$$dC_G = \int_{A_G} dC_G'' dA = -\frac{\varepsilon_0}{z_0^2} W \, dT \qquad (6.7.57)$$

This means that the capacity change is proportional to the volume change. Thus, temperature coefficient TCC is calculated as

$$TCC = -\frac{\varepsilon_0}{z_0^2} \frac{1}{C_{G0}} W = -\frac{W}{z_0 a^2} \qquad (6.7.58)$$

Example 6.25:   Temperature Coefficient of a Micro-GOLAY Cell

We have a GOLAY cell with a silicon base body of a sensor area $4 \times 4 \text{ mm}^2$ and cell depth h of 500 µm. When calculating the volume and size of the bending plate, we have to include the $55°$ inclination of the edges in the silicon body. Edge length a of the bending plate then becomes:

$$a = b - 2(h \tan 55°) = 2.57 \text{ mm} \qquad (6.7.59)$$

with edge length $b = 4 \text{ mm}$ of the sensor area. For the volume, it applies that

$$V_0 = a^2 h + 4\left[a\left(\frac{b-a}{2}\right)\frac{h}{2}\right] = 5.14 \text{ mm}^3 \qquad (6.7.60)$$

With plate thickness $d = 0.2 \text{ µm}$, YOUNG's modulus $E = 330 \text{ MPa}$ and POISSON's ratio $v = 0.3$, the bending strength of the plate according to Equation 6.7.44 becomes

$$D = \frac{3.3 \times 10^8 \frac{\text{N}}{\text{m}^2} (0.2 \text{ µm})^3}{12 \cdot (1 - 0.3)} = 3.14 \times 10^{-13} \text{ Nm} \qquad (6.7.61)$$

For volume change ΔV, it then applies that

$$\Delta V = 2.42 \times 10^{-2} \frac{(2.57 \text{ mm})^6}{3.14 \times 10^{-13} \text{ Nm}} \Delta p = 2.23 \times 10^{-5} \text{m}^5 \, \Delta p \qquad (6.7.62)$$

For the pressure change, it follows according to Equation 6.7.49:

$$\Delta p = \frac{1 \times 10^5 \frac{\text{N}}{\text{m}^2}}{300 \text{ K}} \frac{\Delta T}{\dfrac{1 \times 10^5 \frac{\text{N}}{\text{m}^2}}{5.14 \times 10^{-9}\text{m}^3}\left(2.42 \times 10^{-2}\dfrac{(2.57 \text{ mm})^6}{3.14 \times 10^{-13} \text{ Nm}}\right)} = 7.7 \times 10^{-7} \frac{\text{N}}{\text{m}^2} \frac{\Delta T}{\text{K}}$$

$$(6.7.63)$$

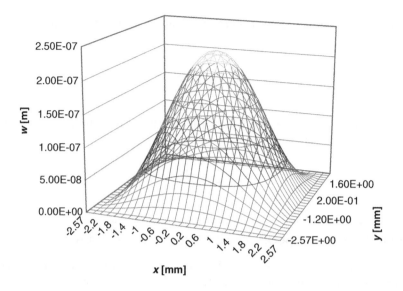

Figure 6.7.4   Deformation of the bending plate at temperature change $\Delta T = 0.1$ K

The deformation of the bending plate can now be calculated using Equation 6.7.43. Figure 6.7.4 shows the deformation of the membrane for a temperature change ΔT of 0.1 K. The maximum deflection of the membrane occurs in the middle ($x = y = 0$):

$$w_{\max} = w(x = 0, \, y = 0) = 228 \text{ nm}.$$

For the individual terms in Equation 6.7.42, it results that

$$\frac{\Delta p}{p_0} = \frac{7.7 \times 10^{-7} \Delta T}{1 \times 10^5} = 7.7 \times 10^{-12} \frac{\Delta T}{K} \qquad (7.6.64)$$

$$\frac{\Delta V}{V_0} = \frac{W \, \Delta T}{5.14 \times 10^{-9}} = 3.33 \times 10^{-3} \frac{\Delta T}{K} \qquad (7.6.65)$$

$$\frac{\Delta T}{T_0} = \frac{\Delta T}{300 \text{ K}} = 3.33 \times 10^{-3} \frac{\Delta T}{K} \qquad (7.6.66)$$

That means that pressure change $\Delta p / p_0$ is negligible and it applies that

$$\Delta V \approx \frac{V_0}{T_0} \Delta T \qquad (6.7.67)$$

Using Equation 6.7.51 we arrive at the same result as it applies that

$$\frac{D}{a^6} \ll \frac{P_0}{V_0} \qquad (6.7.68)$$

and, thus, it results that

$$W = \frac{V_0}{T_0} \qquad (6.7.69)$$

That means that the bending stiffness of the bending plate in practice has no actual influence. The pressure plate has only to be sufficiently elastic to yield to the pressure. Thus, the temperature coefficient becomes

$$TCC = -\frac{V_0}{T_0} \frac{1}{z_0 a^2} \qquad (6.7.70)$$

For this example, the resulting value is

$$TCC = -\frac{5.14 \times 10^{-9} \text{ m}^3}{300 \text{ K}} \frac{1}{5 \,\mu\text{m} \times (2.57 \text{ mm})^2} = -0.518/\text{K} \qquad (6.7.71)$$

This is a very large value. It is put into perspective as actual temperature changes ΔT are usually in the range of only a few mK.

We usually apply a half-bridge circuit to evaluate the capacity change (Figure 6.7.2). With Equations 6.7.16, 6.7.53, 6.7.57 and 6.7.67, the voltage change becomes

$$\Delta V_S = -V_E \frac{C_{G0}}{C_E} \frac{V_0}{T_0 z_0 a^2} \Delta T = V_E \frac{C_{G0}}{C_E} TCC \times \Delta T \qquad (6.7.72)$$

We use Equation 6.2.4 to calculate the voltage responsivity:

$$R_V = \frac{\alpha}{G_{\text{th,G}}} \frac{V_E C_{G0}}{C_E} TCC \qquad (6.7.73)$$

The capacitive read-out states that the dominating electrical noise is the kTC noise according to Equation 4.2.10. Consequently, it applies to NEP:

$$NEP = \sqrt{\frac{k_B T_S}{C_{G0} + C_E}} \frac{G_{\text{th,G}}}{\alpha} \frac{C_E}{V_E C_{G0}} \frac{1}{TCC} \qquad (6.7.74)$$

Example 6.26:   Thermal Resolution of a Micro-GOLAY Sensor

In the following, we will estimate temperature resolution $NETD$ for the micro-GOLAY sensor in Example 6.25. For this purpose, at first we have to calculate thermal conductance G_{th} of the sensor. The heat generated in the absorber is basically dissipated through the gas (air) in the cell to the substrate. The heat flowing into the base body will be neglected. Due to the small distance between bending plate and substrate, we can treat it as a thermal ground. It applies that:

$$G_{th,G} = \frac{\lambda_{Air} a^2}{h} = \frac{0.025 \frac{W}{K \times m} (2.57 \, mm)^2}{0.5 \, mm} = 3.3 \times 10^{-4} \frac{W}{K} \qquad (6.7.75)$$

Capacity C_{G0} amounts according to Equation 6.7.52 to

$$C_{G0} = \varepsilon_0 \frac{a^2}{z_0} = 8.854 \times 10^{-12} \frac{As}{Vm} \frac{(2.567 \, mm)^2}{5 \, \mu m} = 11.7 \, pF \qquad (6.7.76)$$

For the responsivity, it applies – at complete radiation absorption ($\alpha = 1$) and bias voltage $V_E = 1 \, V$ – according to Equation 6.7.70:

$$R_V = \frac{1}{3.3 \times 10^{-4} \frac{W}{K}} \frac{1 \, V \times 11.7 \, pF}{11.7 \, pF} \frac{0.52}{K} = 1571 \frac{V}{W} \qquad (6.7.77)$$

The kTC noise becomes with Equation 4.2.10

$$\tilde{v}_{R,a} = \sqrt{\frac{k_B \times 300 \, K}{2 \times 11.7 \, pF}} = 13.3 \, \mu V \qquad (6.7.78)$$

Thus, the noise-equivalent power becomes

$$NEP = \frac{\tilde{v}_{R,a}}{R_V} = \frac{13.3 \, \mu V}{1571 \frac{V}{W}} = 8.5 \, nW \qquad (6.7.79)$$

NETD of the micro-GOLAY sensor is calculated using Equation 5.4.9 as well as f-number $F = 1$, in the wavelength range of 8–14 μm and at a temperature T_0 of 27 °C. It becomes

$$NETD = \frac{4k^2 + 1}{A_S} \frac{NEP}{I_M} = \frac{5}{(4 \, mm)^2} \frac{8.5 \times 10^{-9} W}{2.64 \, W/(m^2 K)} = 0.001 \, K \qquad (6.7.80)$$

This example shows that micro-GOLAY cells have a very good temperature resolution. A disadvantage, though, is the comparatively complicated structure and limited robustness.

6.8 Comparison of Thermal Sensors

The several types of thermal sensors can only be differentiated by the way they convert temperature change ΔT into an output signal, mainly voltage ΔV. Therefore, the designs of all thermal sensors intend to reach a maximum temperature change of the sensor element. According to Equations 6.2.11 and 6.2.32, this requires a maximum thermal isolation of the sensor elements and, at the same time, minimum thickness of sensor elements in order to achieve a small thermal time constant. All other parameters, such as size and form of the sensor element, housing, and so on, are usually given by the application. Table 6.8.1 provides a comparison of the most important characteristics of the thermal sensors presented in this book.

Table 6.8.1 Comparison of the characteristics of thermal sensors

Principle	Sensor type	Temperature dependence of the output signal	Determining material parameter	Thermal resolution limit	Remarks
Energy converter	Thermoelectric (thermopile)	ΔT	α_S	$\dfrac{\sqrt{M}}{2} BLIP$	M material-dependent
	Pyroelectric	$\dfrac{dT}{dt}$	π_P	$BLIP$	
Parametric converter	Microbolometer	T	$\alpha_B = TCR$	$BLIP$	Requires vacuum
	Bimorph	T	TCC	$BLIP$	Requires vacuum
	GOLAY cell	T	TCC	$BLIP$	

Pyroelectric sensors exclusively detect signals that change over time. Therefore, a main application area is motion detectors where the object to be measured, for example a person, already by its movement generates a changing signal. Further important areas of application of pyroelectric sensors are pyrometers and spectroscopic devices. In pyrometers, the incident radiation is usually mechanically and temporally modulated, for example by a rotating disc (chopper wheel). In this case, the temperature of the chopper wheel is used as reference signal. In spectroscopy, for example in gas analysis devices, we usually use pulsed radiation sources. The advantages of pyroelectric sensors are the simple and inexpensive structure together with high detectivity and excellent long-term stability. The disadvantages are that they require a chopper (which, however, is an advantage in motion detectors) and that they are highly sensitive to shocks (microphony).

Due to the large progress of micromechanical manufacturing procedures, thermoelectric sensors (thermopiles) are inexpensive thermal sensors that are available in large numbers. They do not require a chopper. Their main field of application is contactless temperature measuring at low frequencies. A very popular example is the ear thermometer. An important advantage of thermopiles is the fact that they can be easily integrated using standard technologies of semiconductor manufacturing – which leads to low-cost mass production. A disadvantage is the necessary reference point temperature and the consequently rather complicated signal evaluation. A good temperature resolution always requires many thermocouples connected in series (thermopile). The space required for this sets limits to the miniaturisation of thermoelectric sensors.

The presented microbolometers are exclusively used for imaging sensors in thermography. They do not require chopping either. The advantage of microbolometers is their good compatibility with manufacturing methods of semiconductor standard technologies, the simple spatial integration of the sensor pixels with their respective wiring connection, the simple signal read-out, a good long-term stability and the insensitivity to mechanical vibration. A disadvantage is the vacuum required for operating the pixels.

The bimorphous sensors are a recent development and are still in the laboratory stage. In principle, they have the same potential as microbolometers. However, there is a number of issues such as mechanical noise, long-term stability or microphony, that are not solved yet.

Micro-GOLAY cells are relatively large in volume. They are basically only used for special applications in laboratories, for example the detection of terahertz radiation.

There are several other signal evaluation procedures for thermal sensors, such as measuring the sensor element temperature with a pn-junction [29]. However, they have no practical relevance yet. As almost all physical effects depend on temperature, further options of parametric sensor principles appear to be possible and could be developed.

References

1. Gerlach, G. and Dötzel, W. (2007) *Introduction to Microsystem Technology*, John Wiley & Sons Ltd, Chichester.
2. Erikson, P., Andersson, J.Y. and Stemme, G. (1997) Thermal characterization of surface-micromachined silicon nitride membrans for thermal infrared detectors. *Journal of Microelectromechanical Systems*, **6**(1), 55–61.
3. Graf, A., Arndt, M., Sauer, M. and Gerlach, G. (2007) Review of micromachined thermopiles for infrared detection. *Measurement Science and Technology*, **18**, R59–R75.
4. Völklein, F., Wiegand, A. and Baier, V. (1991) High-sensitivity radiation thermopiles made of Bi-Sb-Te Films. *Sensors and Actuators A*, **29**, 87–91.
5. Ghanem, W. (1999) *Development and Characterization of a Sensor for Human Information: a Contribution to Innovative House Technique; Erlanger Berichte Mikroelektronik*, Shaker Verlag, Aachen.
6. Birkholz, U., Fettig, R. and Rosenzweig, J. (1987) Fast semiconductor thermoelectric devices. *Sensors and Actuators*, **12**, 179–184.
7. PerkinElmar Optoelectronics GmbH, (2001) Remote temperature measurement with PerkinElmer thermopile sensors (pyrometry): A practical guide to quantitative results.
8. Van Herwaarden, A.W. and Sarro, P.M. (1986) Thermal sensors based on the Seebeck effect. *Sensors and Actuators*, **10**, 321–346.
9. Tichý, J. and Gautschi, G. (1980) *Piezoelektrische Meßtechnik (Piezoelectric Measurement Technology)*, Springer-Verlag, Berlin.
10. Moulson, A.J. and Herbert, J.M. (2003) *Electroceramics*, John Wiley & Sons Ltd, Chichester.
11. Ivers-Tiffee, E. and von Münch, W. (2007) *Werkstoffe der Elektrotechnik (Materials in Electrical Engineering)*, Stuttgart, B.G. Teubner Verlag.
12. Kittel, Ch. (2004) *Introduction to Solid State Physics*, 8th edn, John Wiley& Sons Ltd, Chichester.
13. Shvedov, D. (2008) *Beschleunigungsempfindlichkeit Pyrolelektrischer Sensoren (Acceleration Sensitivity of Pyrolelectric Sensors)*, Dissertation, TUDpress, TU Dresden.
14. Norkus, V., Gerlach, G., Shvedov, D. *et al.* (2008) Acceleration sensitivity of pyroelectric single-element detectors based on lithium tantalate. Proc. IRS², AMA Fachverband für Sensorik, pp. 283–287.
15. Neumann, N. and Sänze, H. (2008) How to reduce the microphonic effect in pyroelectric detectors? AMA Fachverband für Sensorik. Proc. IRS², pp. 277–282.
16. Muralt, P. (2005) Micromachined infrared detectors based on pyroelectric thin films, in *Electroceramic-based MEMS: Fabrication-Technology and Applications* (ed. N. Setter), Springer Science + Business Media.
17. Tissot, J.L., Rothan, F., Vedel, C. *et al.* (1998) LETI/LIR's uncooled microbolometer development. *SPIE*, **3436**, 605–610.
18. Kruse, P.W. and Skatrud, D.D. (1997) *Uncooled Infrared Imaging Arrays and Systems; Semiconductors and Semimetals*, vol. **47**, Academic Press, London.
19. Technical Data Package UL03081; ULIS (Frankreich); 05.11.2004.
20. Data sheet U3000AR; Boeing (USA); Mai (1999).
21. Hanson, C.M., Beratan, H.R. and Arbuthnot, D.L. (2008) Uncooled thermal imaging with thin-film ferroelectric detectors. *Proceedings of SPIE*, Vol. 6940, 694025–1–694025–12.
22. Hunter, S.R., Maurer, G., Jiang, L. and Simelgor, G. (2006) High sensitivity uncooled microcantilever infrared imaging arrays. *Proceedings of SPIE*, Vol. 6206, 62061J–1–12.
23. Ivanova, K., Ivanov, Tzv. and Rangelow, I.W. (2005) Micromachined arch-type cantilever as high sensitivity uncooled infrared detector. *Journal of Vacuum Science & Technology*, **B23**(6), 3153–3157.
24. Kwon, I.W., Kim, J.E., Hwang, C.H. *et al.* (2008) A high fill-factor uncooled infrared detector with low noise characteristic. *Proceedings of SPIE*, **6940**, 694014J–1–10.

25. Golay, M.J.E. (1947) A pneumatic infra-red detector. *The Review of Scientific Instruments*, **18**, **5**, 357–362.

26. Yamashita, K., Murata, A. and Okuyama, M. (1998) Miniturised infrared sensor using diaphragm based Golay cell. *Sensors and Actuators A*, **66**, 29–32.

27. Kenny, T.W. (1997) Tunneling infrared sensors, in *Uncooled Infrared Imaging Arrays and Systems; Semiconductors and Semimetals*, Vol. **47** (eds P.W. Kruse and D.D. Skatrud) Academic Press, pp. 227–267.

28. Dym, C.L. and Shames, I.H. (1973) *Solid Mechanics: A Variational Approach*, McGraw-Hill, New York.

29. Ishikawa, T., Ueno, M., Endo, K. *et al.* (1999) Low-cost 320 × 240 uncooled IRFPA using conventional silicon IC process. SPIE, Vol. 3698, 556–564.

30. Logan, R.M. and McLean, T.P. (1973) Analysis of thermal spread in a pyroelectric imaging system. *Infrared Physics*, **3**, 15–24.

31. Erdtmann, M., Zhang, L., Jin, G. *et al.* (2009) Optical readout photomechanical imager: from design to implementation. SPIE Conference on Infrared Technology and Application XXXV, SPIE, Vol. 7298, pp. I-1–I-7.

7

Applications of Thermal Infrared Sensors

7.1 General Considerations

The main application areas of thermal infrared sensors are devices for non-contact temperature measurement (pyrometers and thermal imaging devices), motion detectors, spectrometers and gas analysers. Table 1.1.1 showed the general measuring chains of such systems.

The functional sensor parameters described in Chapters 5 and 6 regarding

- thermal resolution (detectivity D^*),
- spatial resolution (MTF),
- temporal resolution (time constant).

apply to all fields independent of specific applications. The different areas of application pose diverse demands on the sensors. The desired functional parameters and the size parameters that are important for the calculation, for example the sensor area, are the result of the interaction of radiation source, propagation path, imaging system and sensor.

For the description of the radiation source, we usually assume a greybody, that is a blackbody with a wavelength-independent emissivity of $\varepsilon < 1$. These greybodies are usually sufficient for describing natural emitters, if we determine a suitable band emissivity for the selected wavelength range.

Thermal emitters are, in principle, characterised by their temperature and emissivity (see Example 3.4). For non-thermal emitters we have to take into account that other relations than described in Chapters 2 and 5 may apply to the propagation of the radiation on the propagation path as well as to the optical projection. A laser, for instance, emits coherent radiation and the given calculations of optical parameters (Section 5.5) and spatial resolution (MTF; Section 5.6) thus do not apply in the way described there!

Taking into account the radiation source, the transmission of the propagation path determines the wavelength range that can be used. Either the propagation path does not affect the

---

Thermal Infrared Sensors: Theory, Optimisation and Practice Helmut Budzier and Gerald Gerlach
© 2011 John Wiley & Sons, Ltd

measurement (e.g. for temperature measurement) or it includes the measuring object (e.g. for gas analysis).

If we use the atmosphere as propagation path, we can only work with atmospheric windows (Figure 1.1.1). In the optimal case, the maximum of the exitance of the radiation source falls exactly within the selected range (WIEN's displacement law; Equation 2.3.6).

Imaging systems concentrate the radiation to the sensor element (Section 5.5). They often consist of an optics and a filter. For simple images, a single lens is often sufficient. Thermal imaging systems use colour-corrected lens systems (at least two lenses). A filter determines the wavelength range of the measuring system. Spectrometers use gratings and prisms, for instance, for the spectral decomposition of the radiation. Optics and filters may be already integrated into the sensor housing.

For the selection of a sensor, we primarily look at the functional parameters:

- blackbody and spectral responsivity (Section 5.1),
- noise voltage/current (Section 4),
- noise-equivalent power (Section 5.2),
- detectivity (Section 5.3),
- homogeneity of the responsivity over the sensor element (Section 5.1),
- uniformity (Section 5.1),
- time constants (Section 5.1).

In order to calculate the functional parameters we also need some design parameters:

- shape and area of the sensor element,
- aperture angle, field of view (Section 5.5),
- design and size of the entire sensor.

If we have determined the functional and design parameters, we can select temperature resolution *NETD* (Section 5.4) and spatial resolution *MTF* (Section 5.6).

Further important parameters of a sensor are the operational parameters:

- bias current or voltage,
- resistance, capacity,
- type of signal output: analogous or digital,
- operating temperature range,
- thermostatization and the correspondingly required power supply as well as reliability parameters,
- temperature limits for storage and operation,
- life cycle,
- shock, vibration and microphony behaviour.

7.2 Pyrometry

Devices for measuring electromagnetic radiation are called *radiometers* or *radiation meter*. Devices that measure electromagnetic radiation for non-contact measurements of the surface temperature of bodies are called *pyrometers* (greek: *pyr*: fire, *metron*: measure).

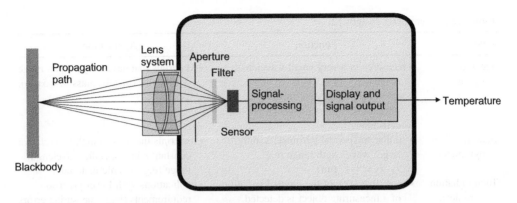

Figure 7.2.1 Schematic set-up of pyrometers.

7.2.1 Design

Figure 7.2.1 shows the structure of a pyrometer.

Usually the devices consist of the structural components presented in Table 7.2.1.

The lens system collects the IR radiation emitted by the measuring object. Interfering marginal rays are blinded out by an aperture. The filter determines the spectral range. As shown

Table 7.2.1 Structural components in pyrometers and their functions [1,2]

Component	Function
Measuring object (radiation source)	• mainly to be considered a greybody (blackbody with band or total emissivity $\varepsilon < 1$, compare Equations 2.2.33 or 2.2.34.
Modulator (propagation path)	• periodically interrupts the optical path (alternating light method, compare Example 6.2.2).
Optics (imaging system)	• concentrates the radiation emitted by the measuring object (compare Figure 5.5.1, Example 5.10), • determines the measuring field (lateral resolution capacity), • affects the thermal resolution method via the field of view, • uses filters to limit the spectral range (Equations 2.25–2.27).
Sensor	• converts the incident (projected) radiation into an electrical signal (mainly thermopiles or pyroelectric sensors, compare Section 6.4 and 6.5).
Sensor signal processing	• amplifies with the lowest noise possible (compare Examples 6.5.3 and 6.5.4 for pyroelectric sensors, amongst other), • implements the best signal-to-noise ratio possible or, respectively, maximum detectivity D^* (e.g., Equation 6.5.78 or 6.5.79), • compensates interferences of ambient temperature fluctuations, • includes emissivity when calculating the value of the measuring temperature (Equations 2.2.32 or 2.2.34), • linearises the output signal, • analogous and/or digital.

Table 7.2.2 Pyrometer designs [2]

Type	Function	Application
Spectral pyrometer	• measures in a very small wavelength range ($\lambda_1 - \lambda_2 \ll \lambda_{1;2}$), • wavelength range determined by interference filters or special photon detectors.	• Materials that have only a small range with high ε (metals, glass: $\lambda \approx 5\,\mu m$).
Band radiation pyrometer	• similar to spectral pyrometers, only larger wavelength range (e.g. $\lambda_1 - \lambda_2 = 8\text{--}13\,\mu m$).	• materials that have a high and relative constant ε in a specific wavelength range (e.g. organic materials).
Total radiation pyrometer	• more that 90% of the emitted radiation of a measuring object is detected, • requires broadband active/transparent lenses, filters and sensors.	• applications with lower precision requirements (large measuring errors due to atmospheric window and $\varepsilon(\lambda)$.
Quotient pyrometer	• measuring the radiant flux at two different wavelengths λ_1 and λ_2 with $\lambda_1 < \lambda_1$ (e.g. 0.95 and 1.05 μm), forming the quotient and calculating the temperature.	• difficult measuring tasks such as high temperatures, obstructed view (smoke, particulate matter), small measuring objects (up to approximately 10% of the measuring field) and unknown emissivities.
Four-colour pyrometer	• simultaneous measurement of the radiant flux in four different spectral ranges, • adaptive emissivity correction with teach-in.	• for small and temporally instable emissivities.

in Table 7.2.2, it is possible to measure within a very small (spectral, quotient pyrometer) or a very large (band radiation pyrometer) wavelength range or to utilise almost the entire spectrum (total radiation pyrometer).

Mainly thermopiles or pyroelectric sensors are used as detectors. Modern microprocessor technology is applied for signal evaluation. Usually RS 232, RS 485 or USB are used as interfaces. However, there are also analogous output signals 4–20 mA and 0–10 V available.

Pyrometers working according to the alternating light principle have a mechanical modulator in the optical path, usually a swinging or a multibladed chopper (see Figure 6.1.1). The optical path is thus periodically interrupted and the sensor measures the difference between object radiation and (thermostatable) chopper, on the one hand, and the sensor signal with minimum bandwidth, on the other hand. Both improve detectivity and, consequently, result in smaller *NETD* values.

Pyrometers can be equipped with additional components, such as

- sighting device (through-lens view finder or laser pointer, that indicates the measuring field and its centre),
- blowing cap, that protects the optics from dust or particulate matter and from fogging up due to condensate,
- cooling devices (radiation plate, water-cooled plates or cooling housing).

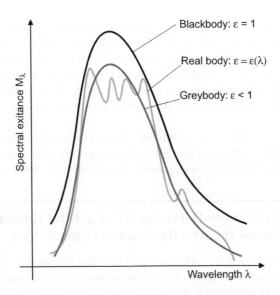

Figure 7.2.2 Spectral exitance of blackbody, greybody and real emitters with the same temperature.

7.2.2 Emissivity of Real Emitters

Real emitters can only be considered approximations of greybodies (Figure 7.2.2).

The more the emissivity can be considered to be constant in at least some ranges, the better the approximation (Table 7.2.3).

In order to measure the surface temperature of a measuring object with a pyrometer, we have to know emissivity ε of the measuring object. Typically, we enter the emissivity and the ambient

Table 7.2.3 Emissivity ε of technical surfaces [2]

	Material	Spectral range 1.4–1.8μm	Spectral range 4.9–5.5 μm	Spectral range 8–14 μm
Metal	steel, polished	0.30–0.40	0.10–0.30	0.10–0.30
	steel, rolled	0.35–0.50	0.20–0.30	0.20–0.30
	steel, oxidised	0.80–0.90	0.70–0.90	0.60–0.80
	copper, polished	0.06–0.20	0.05–0.10	0.03–0.10
	copper, oxidised	0.40–0.80	0.20–0.70	0.20–0.70
	aluminium, polished	0.05–0.25	0.03–0.15	0.02–0.15
	aluminium, anodised	0.10–0.40	0.10–0.40	0.95
Non-metals	coal, graphite	0.70–0.95	0.70–0.95	0.70–0.95
	stone, earth, ceramics	0.40–0.70	0.50–0.80	0.60–0.95
	lacquers, paints	—	0.60–0.90	0.70–0.95
	woods, plastics, paper	—	0.60–0.90	0.80–0.95
	textiles	0.70–0.85	0.70–0.95	0.75–0.95
	thin glass	0.05–0.20	0.70–0.95	0.75–0.95
	water, snow, ice	—	—	0.90–0.95

temperature of the measuring object into the pyrometer. Both values are then used for calculating the temperature during sensor signal processing. The emissivity of metals is problematic. For polished surfaces, it is low and decreases with increasing wavelength. Oxidised or worn surfaces have considerably higher emissivities that can be highly dependent on temperature and wavelength though. Shining metal surfaces reflect strongly and can change the measuring conditions. Thus, they constitute the most difficult measuring objects in pyrometry.

Glasses are transparent in the range of smaller wavelengths of up to 3 μm, whereas they show a high emissivity between 4 μm and 9 μm. In practice, we mainly use wavelengths around 5 μm, as there the atmosphere has a high transparency and glass hardly reflects.

Example 7.1: Systematic Deviation ΔT of a Total Radiation Measurement for an Incorrectly Assumed Emissivity ε

We will analyse how large systematic deviation ΔT of the indicated temperature becomes for a pyrometric measurement where the pyrometer was calibrated using a blackbody and emissivity $\varepsilon = \varepsilon_0 + \Delta\varepsilon$ deviates by $\Delta\varepsilon$.

For a total radiation pyrometer (index T), radiant flux M relating to an area and being projected onto the sensor of the pyrometer becomes according to Equation 2.3.8

$$M = \varepsilon_T \sigma T_T^4 \tag{7.2.1}$$

For the same radiant flux, it results for correctly (ε_T) and incorrectly ($\varepsilon_T + \Delta\varepsilon$) assumed emissivity

$$\varepsilon_T \sigma T_T^4 = (\varepsilon_T + \Delta\varepsilon)\sigma(T_T + \Delta T_T)^4 \tag{7.2.2}$$

After transformation, it results

$$\Delta T_T = \left(\frac{1}{\sqrt[4]{1 + \dfrac{\Delta\varepsilon}{\varepsilon_T}}} - 1 \right) T_T \tag{7.2.3}$$

With the approximation

$$\frac{1}{(a+x)^n} \approx \frac{1}{a^n} - \frac{nx}{a^{n+1}}$$

from the TAYLOR series presentation for $x = a$, it follows that

$$\Delta T_T \approx -\frac{1}{4}\frac{\Delta\varepsilon}{\varepsilon_T} T_T \tag{7.2.4}$$

Figure 7.2.3 shows for total radiation pyrometers the effect of incorrect emissivity values $\Delta\varepsilon/\varepsilon_T$ on the calculated temperature.

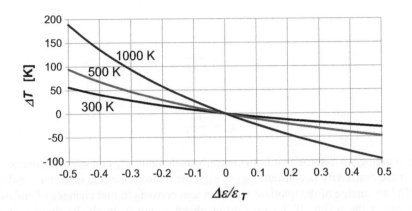

Figure 7.2.3 Systematic deviation ΔT for total radiation pyrometers due to incorrectly assumed emissivities. Parameter: object temperature T_T.

Example 7.2: Temperature Deviation of a Quotient Pyrometer

If the radiation power is sufficiently high, we can measure in very small spectral ranges that can be assumed to have constant emissivity ε_S (index S for spectral). Wavelength range $\lambda_1 \ldots \lambda_2$ is substantially smaller than medium wavelength $\lambda_S(\lambda_2 - \lambda_1 \ll \lambda_S = (\lambda_1 + \lambda_2)/2)$. Again we start from Equation 2.3.7. With the approximation $\exp\{c_2/(\lambda_S T)\} \gg 1$ (error smaller than 1%), it follows for the exitance

$$M \approx \varepsilon_S \int_{\lambda_1}^{\lambda_2} \frac{c_1}{\lambda^5} \frac{1}{\exp\left\{\dfrac{c_2}{\lambda T}\right\}} \, d\lambda \tag{7.2.5}$$

With the integral calculus, such as used in Example 2.3.1, applying Equation 2.3.17, the specific exitance becomes

$$M = \varepsilon_S \frac{c_1 T^4}{c_2^4} \exp\{-x\} \left(x^3 + 3x^2 + 6x + 6\right) \tag{7.2.6}$$

with

$$x = \frac{c_2}{\lambda_S T} \tag{7.2.7}$$

Now we determine the quotient of both specific exitances:

$$\frac{M_1}{M_2} = \frac{\varepsilon_{S1} \exp(-x_1)[(x_1)^3 + 3(x_1)^2 + 6(x_1) + 6]}{\varepsilon_{S2} \exp(-x_2)[(x_2)^3 + 3(x_2)^2 + 6(x_2) + 6]} \approx \frac{\varepsilon_{S1} \exp(-x_1)}{\varepsilon_{S2} \exp(-x_2)} \tag{7.2.8}$$

with

$$x_1 = \frac{c_2}{\lambda_{S1} T}$$

(7.2.9)

and

$$x_2 = \frac{c_2}{\lambda_{S2} T}$$

(7.2.10)

For rounding we assume that both central wavelengths λ_{S1} and λ_{S2} are close to each other. If the emissivity is constant $(\varepsilon_{S1} = \varepsilon_{S2})$, the measurement is emissivity-independent. The advantage of the quotient measurement consists in that changes of emissivity do not affect the result. If the measuring object is not a greybody, there will be a measuring deviation:

$$\Delta\left(\frac{M_1}{M_2}\right) = \frac{\varepsilon_{S1} \exp(-x_1)}{\varepsilon_{S2} \exp(-x_2)} - \frac{\exp(-x_1)}{\exp(-x_2)}$$

(7.2.11)

Setting Equation 7.2.11 to zero we can calculate temperature difference $\Delta T = T_M - T$ that presents the systematic measuring deviation with T_M being the measured temperature and T the actual temperature:

$$\frac{\varepsilon_{S1}}{\varepsilon_{S2}} \exp\left\{\frac{\lambda_{S1} - \lambda_{S2}}{\lambda_{S1}\lambda_{S2}} \frac{c_2}{T_M}\right\} = \exp\left\{\frac{\lambda_{S1} - \lambda_{S2}}{\lambda_{S1}\lambda_{S2}} \frac{c_2}{T}\right\}$$

(7.2.12)

A simple transformation results in

$$\left(\frac{1}{T} - \frac{1}{T_M}\right) = \frac{1}{c_2} \frac{\lambda_{S1}\lambda_{S2}}{\lambda_{S1} - \lambda_{S2}} \ln \frac{\varepsilon_{S1}}{\varepsilon_{S2}}$$

(7.2.13)

Applying

$$\frac{1}{T} - \frac{1}{T_M} = \frac{T_M - T}{T T_M} = \frac{\Delta T}{T} \frac{1}{T_M}$$

(7.2.14)

the temperature deviation becomes

$$\frac{\Delta T}{T} = \frac{T_M}{c_2} \frac{\lambda_{S1}\lambda_{S2}}{\lambda_{S1} - \lambda_{S2}} \ln \frac{\varepsilon_{S1}}{\varepsilon_{S2}}$$

(7.2.15)

7.3 Thermal Imaging Cameras

Thermal imaging cameras (also called thermal camera, thermography system, infrared camera) are imaging systems that approximate the temperature distribution of a scene. They

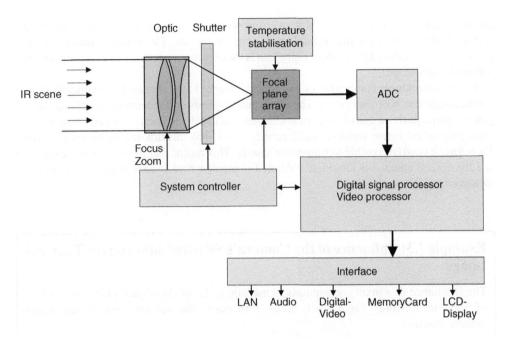

Figure 7.3.1   Design of a thermal imaging device

represent the radiation emitted by a body or object in a two-dimensional image (thermal image, infrared image). The radiation emitted by the object is composed of the object's own radiation, the reflected ambient radiation and possibly also of the transmitted background radiation. Therefore, the thermal image only constitutes an approximation of the temperature distribution in a scene.

Thermal imaging cameras with thermal infrared sensors are also called uncooled thermal imaging cameras. They are mainly used for measuring tasks in the ambient temperature range and thus usually operate in the atmospheric window of the far-infrared range (approximately $\lambda = 8–13\,\mu m$). In the mid-infrared range ($\lambda = 3–5\,\mu m$), uncooled cameras are only used for specific applications, mainly at higher temperatures [8].

The operating principle of a thermal imaging camera corresponds to that of a conventional video camera in the visible spectral range (Figure 7.3.1) where each individual pixel can be assumed to be a pyrometer (Section 7.2). Therefore, in the following, we will only describe selected, sensor-array-related details.

7.3.1   Design

Figure 7.3.1 shows the principal structure of thermal imaging cameras. The infrared optics projects the measuring scene onto the image sensor. The lenses of the optics are made of germanium, silicon or special glasses with high transmissivity in the far-infrared range. For uncooled cameras, we usually use high-aperture optics (f-number $F < 1.4$). Automatic focussing and zoom optics are the technological state of the art. The smallest feasible spatial resolutions are produced at a scale of unity, that is the object is projected 1 : 1 to the sensor array (infrared microscope).

A system-imminent component of the optical channel is the shutter. It is cyclically closed for the recalibration of the system. The cycle time depends on the temperature change of the camera, particularly on that of the optical channel, and usually amounts to several minutes. The shuttering takes a few hundred milliseconds. This may disturb the measuring process. Therefore, it is often possible to trigger the shutter. This means that the system is recalibrated and the optical channel is closed at a defined moment where it does not interfere with the measurement (Example 7.3).

Example 7.3: Influence of the Camera's Self-Radiation on the Thermal Image

The irradiance of a pixel corresponds to the sum of L_O of the object's radiance and L_C of the surrounding (Figure 6.1.3). In this case, the environment is the inner camera housing:

$$E = L_O \omega_{FOV} + L_C \omega_C \tag{7.3.1}$$

with the reduced solid angle of the camera housing radiation

$$\omega_C = \pi - \omega_{FOV} \tag{7.3.2}$$

In order to prevent reflections in the optical channel, the interior of the camera is blackened. Thus, the camera can be assumed to be a blackbody and radiance becomes

$$L_C = \frac{\sigma}{\pi} T_C^4 \tag{7.3.3}$$

If radiance L_O of the object is constant, but interior temperature T_C of the camera changes, each pixel experiences a changed irradiance of

$$\Delta E = \Delta L_C \omega_C \tag{7.3.4}$$

As each pixel has a different reduced solid angle ω_{FOV} (Section 3.2.2.2), the irradiance change of each pixel is also different. It follows from Equations 7.3.2 and 3.2.40:

$$\omega_C = \frac{\pi}{2} \left(1 + \frac{a^2 + h^2 - r^2}{\sqrt{(a^2 + h^2 + r^2)^2 - 4r^2 a^2}} \right) \tag{7.3.5}$$

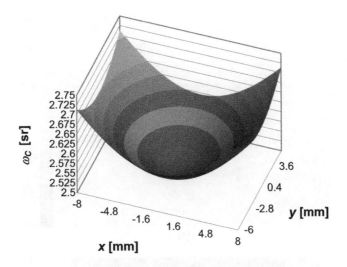

Figure 7.3.2   Reduced solid angle ω_C of the camera housing radiation. f-number $F = 1$; $r = 9$ mm; $h = 18$ mm; array sensor 16×12 mm^2.

with a being the distance of the individual pixel to the optical centre of the array. Even for very high aperture optics with an f-number of $F = 1$, the reduced solid angle of the camera radiation is much larger than that of the scene ($\omega_{FOV} = 0.62$ sr; $\omega_C = 2.52$ sr; see Figures 3.2.8 and 3.2.9). This way even small changes of the camera temperature have a substantial effect on the thermal image (Figure 7.3.2). Whereas the reduced solid angle for the pixel in the array centre ($x = y = 0$ mm) amounts to 2.52 sr, it is 2.71 sr for the farthest pixel ($x = \pm 8$ mm; $y = \pm 6$ mm).

As in addition to the optical channel, the pixel sees different objects – such as the optics edge and sensor housing – and the camera does not evenly warm, Equation 7.3.4 is not sufficient for calculating the correction of the camera's internal radiation. Therefore, the optics is temporarily covered with a shutter whose temperature is known. Based on the recorded thermal images, we can calculate correction factors for each individual pixel.

Figure 7.3.3a shows the thermal image at closed shutter directly after the shutter correction. The indicated temperature approximately corresponds to the camera's internal temperature ($T_C = 39.0\,°C$). Figure 7.3.3b was recorded about 30 minutes after Figure 7.3.3a, with the temperature having changed ($T_C = 40.8\,°C$). We can clearly see the uneven change of the signal in Figure 7.3.2. Now the indicated temperature values do not coincide with the camera's internal temperature as the adjustment of the camera refers to *FOV* and the individual pixels now receive radiation from the entire half-space. Looking with the thermal imaging camera at a blackbody, the result is the temperature curve presented in Figure 7.3.4. We can also clearly discern the drifting of the pixel values. The fluctuations of the temperature curve corresponds to the noise of the pixel and thus of *NETD* (about 100 mK). The temporarily large measuring values at time t_s of the shutter

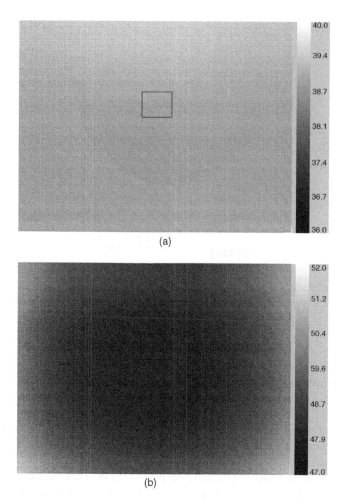

Figure 7.3.3   Thermal images of a warmed camera (a) shortly after shuttering and (b) 30 minutes later

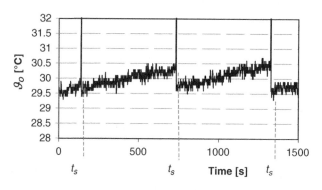

Figure 7.3.4   Curve with temperature values of a pixel in relation to time. t_s shutter time

---

(shutter closed) are caused by the shutter temperature (about 41 °C) and are usually blinded out.

The temperature deviation of a camera is therefore represented by two values (values of technologically top-range devices):

- *NETD*: statistical measuring deviation within one image, for example 50 mK;
- absolute temperature deviation: permissible maximum measuring deviation of the temperature values in respect to the actual temperature in the measuring scene, for example $1\,K \pm 1\%$.

Currently, commercially available thermal imaging cameras exclusively use microbolometer arrays (Section 6.6). Thermoelectric imaging sensors are currently in the laboratory stage [9]. Even though the development of pyroelectric array sensors is very advanced [10], they have been only used in a very limited number of applications. Due to the required modulation of the radiation with a chopper, changes of the camera temperature are not visible in the image (Example 7.4). However, the chopper – that is a moveable mechanical part – is probably the reason why it is hard to establish pyroelectric matrices on the market. Line sensors (1D arrays, line arrays) [11] that are used in spectrography, for instance, are obvious examples of the large potential that pyroelectric arrays have.

Example 7.4: Spiral Choppers

The temporal modulation of the radiant flux with a rotating multi-bladed chopper results in an almost rectangular time curve (Figure 6.2.6). At the end of a chopper phase, the signal reaches its maximum and has to be sampled. By calculating the difference between the chopper-open and chopper-closed phase we arrive at the actual signal ΔV_S. In order for all pixels to be sampled at the optimum time, the chopper phase has to be synchronous to the read-out time of the integrated circuit via the sensor chip (Figure 6.6.9). Specially designed choppers almost achieve this [12]. A sufficient solution is the ARCHIMEDIAN spiral (spiral chopper) which allows the equidistant choppering of the individual rows of the array:

$$r = \alpha B + d \qquad (7.3.6)$$

with angle $\alpha = (0 \ldots 2\pi)$ and

$$B = \frac{h}{2\pi} \qquad (7.3.7)$$

with array height h. Distance d represents the distance of the sensitive area of the array to the chopper centre. The spiral chopper presented in Figure 7.3.5 realises one chopper-open and one chopper-closed phase per revolution. It is balanced and therefore not circular.

The corresponding apertures ensure that only the radiant flux of the object is modulated. Therefore, the pixel does not see any modulated camera housing radiation; and no correction with shutters (Example 7.3) is required.

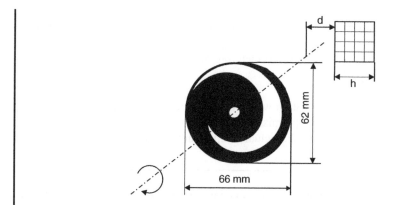

Figure 7.3.5   Spiral chopper of a pyroelectric array sensor with a sensor area of $10 \times 10 \, \text{mm}^2$

Following the analogue-to-digital conversion of the sensor signal with 14–16 bits, the image is processed using signal processors and is then available to the user.

We distinguish between two structural designs of thermal imaging cameras:

- *Portable cameras* include all the necessary controls, storage capacity for the IR images and different interfaces (Figure 7.3.1). These are universal infrared cameras for general use. Modern thermal imaging systems often include a second camera in the visible spectral range, which is used for an improved visualisation of the measuring results, but not for temperature measurements. In addition to the already mentioned visual camera, for the documentation and evaluation of the measuring results, portable cameras often also include an audio interface for recording comments and a laser indication for marking the measuring point.
- *Stationary systems* are designed for industrial applications (integrated systems). They usually only have one interface, for example Ethernet, and do not have any controls directly on the housing. The design of the camera housing is adapted to the specific application. This way it is possible to realise the required safety levels.

7.3.2   Calibration of Thermal Imaging Cameras

During the calibration of a thermal imaging camera, the deficiencies of the sensor (Section 5.1.5) have to be corrected. The calibration requires several steps:

- correction of the characteristic curve,
- replacement of dead pixels,
- temperature calibration.

Table 7.3.1 presents the typical adjustment sequence of an uncooled thermal imaging camera.

Table 7.3.1 Calibration sequence for a thermal imaging camera

Step	Activity
1	Identifying and correcting dead pixels by detecting an incorrect operating point
2	Establishing the V/T_O curve for each pixel
3	Calculating the pixel-specific correction coefficient
4	Measuring the relation of the normal curve to the internal camera temperature
5	Calculating the correction coefficient for the internal temperature
6	Identifying further dead pixels in the corrected thermal image for which responsivity and noise lie outside a defined tolerance; and other interferences
7	Radiometric calibration: identifying a precise signal-temperature curve of the normal curve
8	Integrating the correction algorithms/coefficients into the signal processing of the system

7.3.2.1 Non-Uniformity Correction

Sensor arrays have an instance- and type-specific spatial distribution of both the operating points (offsets) and the responsivity. This is mainly caused by the technology-related parameter fluctuations between pixels. In addition, there are differences due to the optical imaging, especially the different fields of view of the pixels.

This difference in responsivity has to be compensated by pixel-specific signal corrections in order to achieve – for a homogeneous radiation of the sensor array – a uniform output signal ('smoothing out'). For this, we can use several mathematical methods which differ particularly regarding the number of degrees of freedom for the correction. In the following, we will present a simple way to correct the curve [13,14], for instance, to include further examples.

At first we have to transform all pixel curves to a uniform curve, the so-called normal curve. This means, that we normalise the operating points (offsets) and the slope (responsivity) of the curves of the individual pixels. The mean value of all N pixels at object temperature T_O is suitable as a normal curve:

$$V_{normal}(T_O) = \frac{1}{N}\sum_{i=1}^{N} V_i(T_O)\tag{7.3.8}$$

The correction of pixel values $V_i(T)$

$$V_{corr,i}(T_O) = [V_i(T_O) - offset_i] \times gain_i\tag{7.3.9}$$

uses the two correction factors (Figure 7.3.6)

$$offset_i = \frac{V_i(T_2) \times V_{normal}(T_1) - V_i(T_1) \times V_{normal}(T_2)}{V_{normal}(T_2) - V_{normal}(T_2)}\tag{7.3.10}$$

$$gain_i = \frac{V_{normal}(T_1)}{V_i(T_1) - offset_i}\tag{7.3.11}$$

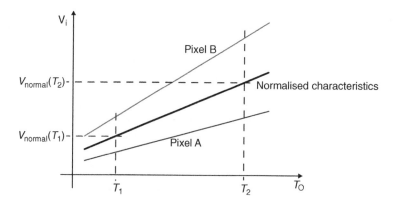

Figure 7.3.6   Variables for the correction of the curve

The value *offset$_i$* is a pure calculatory variable for the correction and does not constitute any measure of the curve difference at a particular temperature.

The presented correction algorithm requires a linear characteristic curve. This means that two points of a straight line are sufficient for calculating the coefficient (2-point correction). For non-linear curves, there are deviations between the support points. In this case, the curve can be described with a second-degree polynomial (3-point correction).

The last step of the curve correction is to determine the dependence of the normal curve on the camera's internal temperature. Shutters are used to compensate the pixel-related changes. This means that we can apply a 2-point correction of the normal curve in order to correct the internal camera temperature.

7.3.2.2 Radiometric Standardisation

After completing the curve correction we have to determine the relation of normal curve and object temperature. For this purpose, we record an object temperature-signal voltage curve. Usually, this curve can be described by a function based on PLANCK'S law [15]:

$$V(T_O) = \frac{R}{e^{\frac{B}{T_O}} - F} + O \qquad (7.3.12)$$

with B, F, O and R being constants that can be determined by measuring the curve. Constant R can be used to adjust the system response. Constant B corresponds – according to PLANCK'S radiation law – theoretically to c_2/λ and describes the spectral behaviour, that is the system's effective wavelength. Constant F will adjust for the non-linearity of the system – for an ideal linearity between radiation and signal, it is One. We have to introduce offset O in order to take into account the system's operating point. At least five measured V/T_O value pairs are necessary to determine the four coefficients B, F, O and R via non-linear correction calculation.

Figure 7.3.7 provides an example.

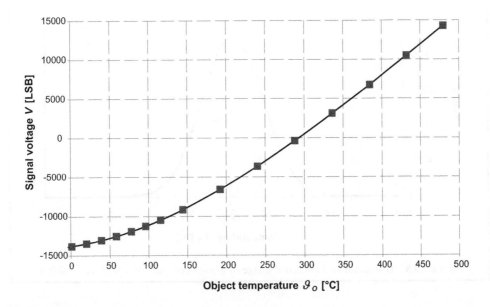

Figure 7.3.7   Object temperature-vs-signal voltage curve of a thermal imaging camera with micro-bolometer. Points: measuring values; full line: regression curve according to Equation 7.3.12 with $B = 1575.3$ K; $F = -0.11\ 173$; $O = -14\ 560$ and $R = 236\ 840$

Example 7.5: Inhomogeneity-Equivalent Temperature Difference

Despite correcting the pixel curves, there remains still a small deviation between the individual pixels. The reasons are uncertainties when determining the correction factor, caused by pixel noise, non-linear curves and deficiencies of the measuring device, for instance. Even though these deviations are determined, for arbitrary (random) scenes, the observer perceives them as noise (see Section 5.1.5) and they are therefore interpreted as spatial noise.

Replacing the noise voltage in the *NETD* calculation (Equation 5.4.4) with the pixel voltage scattering of a homogeneous irradiation of all pixels, we arrive at the inhomogeneity-equivalent temperature difference *IETD*.

Figure 7.3.8 presents the histogram of a thermal image of a homogeneous scene with 30.0 °C. This figure is based on the array in Figure 5.1.7 (histogram of the operational point voltages) and was used for describing the adjustment. We now can use the approximately GAUSS-distributed curve to directly read out the scattering: *IETD* = 80 mK.

Total *NETD* is then calculated by the squared sum of the temporal and spatial noise-equivalent temperature resolutions:

$$NETD_{total} = \sqrt{NETD^2 + IETD^2} \qquad (7.3.13)$$

NETD was calculated at 100 mK in the example system. Thus, it applies that $NETD_{total} = 128$ mK. The absolute measuring deviation is 0.88 K.

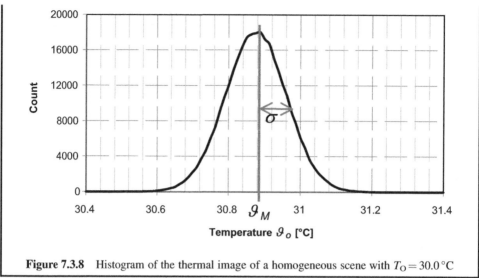

Figure 7.3.8 Histogram of the thermal image of a homogeneous scene with $T_O = 30.0\,°C$

7.4 Passive Infrared Motion Detector

Motion detectors are used for switching specific devices if there are people present. This may be a light that is switched on if somebody is approaching a certain area, or an alarm that is triggered at the approach of an unauthorised person. Motion detectors detect the heat radiation of people who according to PLANCK's radiation law in Equation 2.3.3 and Figure 3.2.1 in the body temperature range reach a maximum irradiation at a wavelength of approximately 10 μm. As the people themselves are the radiation source, such detectors are called passive infrared motion detectors or short PIR motion detectors [3].

7.4.1 Design

Typically motion detectors have the structure shown in Figure 7.4.1 (Table 7.4.1).

The IR radiation of a person that enters the propagation path of the motion detector (area between the projected and the actual sensor elements of the dual sensor) is projected through a lens to the pyroelectric sensor and causes an electrical sensor output signal V_S. Usually, motion detectors use pyroelectric detectors as – according to Equation 6.5.18 and due to their working principle – they generate a sensor output signal that is not proportional to the temperature, but proportional to the temporal change of temperature $d(\Delta T_S)/dt$. In this context, this is of particular interest as the goal is to detect people. Usually, dual sensors are used, with the single detectors having opposite poles. This means that when the projection transits from one sensor to the other a twice as large sensor signal is generated.

Detection is only possible in the x-direction of the coordinate system presented in Figure 7.4.1. If the person moves in y- or z-direction along the main axis of the coordinate system the radiant flux on the sensor elements does not change. This means, the person

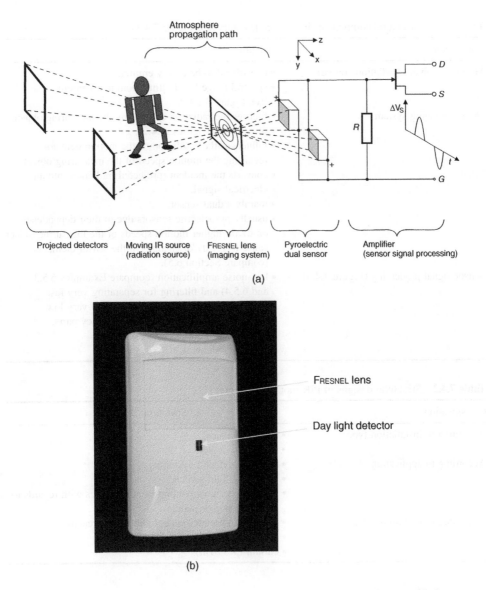

(a)

(b)

Figure 7.4.1   Design of PIR motion detectors. (a) schematic set-up, (b) photographic image

cannot be detected. As motion detectors should detect all movements within their detection area they are usually mounted with a twist. For a central approach, the IR ray covers both elements of the dual sensor both for tangential and central approaching and generates an output signal that can be evaluated. Due to the operating principle, a tangential approach leads to a significantly larger response signal (in practice, often twice as large as for a central approach).

Table 7.4.2 summarizes the classification of PIR motion detectors.

Table 7.4.1 Structural components in motion detectors (see Figure 7.4.1)

Component	Function
Measuring object (radiation source)	• considered to be a grey emitter, • spectral range 8…13 µm (room temperature [see Figure 3.2.1]).
Optics (imaging system)	• focuses the radiation emitted by the measuring object, • often FRESNEL lenses (7.4.2), • influences the resolution of the motion detector regarding the motion speed of the measuring object.
Sensor	• converts the incident (projected) radiation into an electrical signal, • mainly a dual sensor, • usually pyroelectric sensors due to their dependence on dT/dt: higher motion speeds of the measuring object cause larger dT/dt changes in the sensor elements (compare Section 6.5).
Sensor signal processing (Figure 7.4.3)	• low-noise amplification (compare Examples 6.5.3 and 6.5.4) and filtering for separating very fast (interference of passing vehicles) and very low (influence of sun and clouds) frequency parts, • threshold switch, • switching element.

Table 7.4.2 Structural designs of PIR motion detectors

Classification	Structural Design
According to installation type	• surface devices, • concealed devices.
According to application	• intrusion detector systems, • automatic light switch, • presence detectors (monitoring of spaces with regards to the presence of people).
According to monitoring area	• volumetric (detection angle 86°, 12 m reach), • long-distance (7°, 20 m), • curtain (5°, 10 m).

7.4.2   Infrared Optics

The output signal $\Delta V_S(t)$ of the pyroelectric dual sensor becomes larger:

• for larger radiant flux differences $\Delta\Phi(t)$ between the respective sensitive areas of the sensor,
• for larger temporal changes $d(\Delta\Phi(t))/dt$ of the radiant flux.

The following requirements regarding the IR optics can be derived:

• As motion detectors are usually supposed to detect people and objects in the ambient temperature range, according to Figure 3.2.1, the optics have to have a large transmittance in the wavelength range around 10 µm.

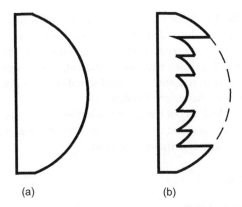

(a)   (b)

Figure 7.4.2   (a) Collecting lens and (b) FRESNEL lens with the same focal length

- In order to achieve a high responsivity regarding different view angles in a large solid angle range with a single motion detector, there has to be separate elementary lenses for as many spatial directions as possible. The arrangement of these individual lenses determines the detection area of the motion detector. If we want to cover a larger horizontal motion range, several rows of lenses have to be mounted above each other.
- The temporal change of sensor output signal $d(\Delta V_S)/dt$ can be increased if the projection of the optical imaging systems leads to large spatial temperature gradients. The arrangement of a large number of individual lenses helps this purpose.

Plano-convex lenses with a large 'collecting area', that is a large diameter, that project the IR-radiation by refraction onto the sensor elements would be rather thick, and thus bulky and heavy. Thick plastic lenses, for instance, would already absorb a large part of the heat radiation and, consequently, weaken the radiant flux incident on the sensor. Step or FRESNEL lenses are an alternative (Figure 7.4.2). Typical thicknesses of FRESNEL lenses in PIR motion detectors are 0.4 mm. They can be easily manufactured by imprinting plastic foils or injection moulding.

7.4.3   Signal Processing

Figure 7.4.3 shows the typical block diagram of an automatic light switch.

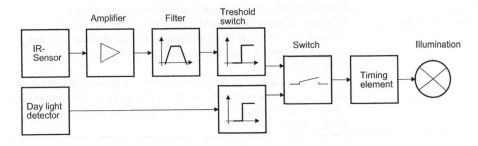

Figure 7.4.3   Block diagram of an automatic light switch

The electric signal of the IR sensor is amplified and filtered with a bandpass. The filtering will prevent the triggering of undesired switching processes by too high-frequency (too fast processes, such as passing cars) or too low-frequency parts (too slow processes, such as sun and clouds). The threshold switch ensures that only sufficiently large signals from people in the detection area trigger the switch, whereas undesired signals from animals or windswept leaves that are usually substantially smaller are blinded out.

As during daylight the light does not have to be switched on, we also need a daylight sensor (light-sensitive switch). A comparison of the daylight sensor signal with a defined threshold value determines whether there is sufficient daylight.

7.5 Infrared Spectrometry

7.5.1 Radiation Absorption of Gases

Atoms and molecules show a large variety of interaction with electromagnetic radiation. This refers to all frequency and wavelength ranges (Table 7.5.1).

Table 7.5.2 shows the fundamental oscillation of simple molecules. If a molecule has N atoms, the position of each atom is determined by three coordinates x, y and z. The molecule has $3N$ degrees of freedom (DOF). The $3N$ degrees of freedom determine the bonding distances and angles between the atoms of the molecule. The molecule itself can move in space in three directions and can turn around three axes (for linear molecules around two axes as the longitudinal axis is the symmetry axis). This means that a total of $3N–5$ (for linear) or $3N–6$ (for non-linear molecules) degrees of freedom remain. Each DOF corresponds to a possible fundamental oscillation.

Table 7.5.1 Interaction of atoms and molecules with electromagnetic radiation

Wavelength range	Effect
Radio frequency range 100 m – 1 cm	Electron spin causes a very small magnetic dipole; inverting the spin causes a change of the dipole's spatial direction that interacts with the magnetic part of the radiation and leads to absorption or emission.
Microwave range 1 cm – 10 µm	Molecules with permanent dipole moment align according to the electric component of the radiation (rotation); this leads to a wavelength-dependent absorption or emission.
IR range 100 µm – 0.78 µm	Oscillations of the atoms in the molecules that cause a change of the dipole moment interact with the electric component of the radiation (IR-active). If the dipole moment in the molecule remains constant for specific oscillation modes, it is IR-inactive (see Table 7.5.2).
VIS/UV range 0.78 µm – 10 nm	Alternating electric field of the radiation induces a periodic deflection of the electrons in the molecules and thus a change in the dipole moment; it causes the absorption of the radiation at resonance frequency of the deflection.

Table 7.5.2 Fundamental oscillations of simple molecules (according to [4])

Molecule type	Example		IR-active	Wavenumber v (cm^{-1})
Linear molecules $DOF = 3N-5$	Carbon dioxide CO_2 $N=3\ DOF=4$	Symmetric valence oscillation	no	1330
		Anti-symmetric valence oscillation	yes	2349
		Deformation oscillation	yes	667.3
		Deformation oscillation perpendicular to the blade plane	yes	667.3
Non-linear molecules $DOF = 3N-6$	Water H_2O $N=3;\ DOF=3$	Symmetric valence oscillation, parallel	yes	3652
		Anti-symmetric valence oscillation, perpendicular	yes	3756
		Symmetric deformation oscillation	yes	1593

Fundamental oscillations of linear molecules that do not generate a dipole moment, for example symmetric valence oscillations of CO_2, are IR-inactive; the others absorb at the respective resonance frequency or resonance wave number energy and are IR-active.

The condition that the dipole moment has to change is a limiting factor for IR spectroscopy. However, the fact that certain oscillations can be observed or not, allows conclusions regarding the structure of certain molecules.

At IR spectrometry, the different oscillations occur simultaneously – not only the harmonic oscillations in Table 7.5.2, but also the higher harmonics, that is the multiples of theses frequency with intensity decreasing with increasing frequency. In addition, the oscillations affect each other causing combination oscillations.

Due to $3N–6$ or $3N–5$ degrees of freedom, the infrared spectrum of large molecules shows numerous normal oscillations that can be divided into two classes:

- Skeletal oscillations: The oscillations comprise many atoms of the molecule. For organic molecules, it usually lies in the wavelength range of $1400–700 \, cm^{-1}$.
- Group oscillations: They concern only a small part of the molecule, mainly certain molecular groups (Table 7.5.3). The rest of the molecule remains in a more or less quiescent state. It is common in spring-mass systems that resonance frequency or wavenumber v becomes the larger, the lighter the oscillating atoms in the terminal groups are. Due to their small

Table 7.5.3   Characteristic valence oscillation wave numbers of different molecular groups (according to [4])

Group	Wave number[a] (cm^{-1})	Group	Wavenumber[a] (cm^{-1})
$-OH$	3650	$>C=O$	1720, 1780[b]
$-NH_2$	3400	$>C=O<$	1650
$\equiv CH$	3300	$>C=N<$	1600
⬡–H	3060	⬠N–	1350
$=CH_2$	3030	$-C-C-$	1200–1000
$-CH_3$	2970, 2870, 1460, 1375[b]	$-C-N<$	1200–1000
$-CH_2-$	2930, 2860, 1470[b]	$-C-O<$	1200–1000
$-SH$	2580	$>C=S$	1100
$-C\equiv N$	2250	$-C-F$	1050
$-C\equiv C-$	2220		

[a]Approximate value.
[b]According to oscillation type (symmetrical or asymmetrical, stretching or deformation oscillation).

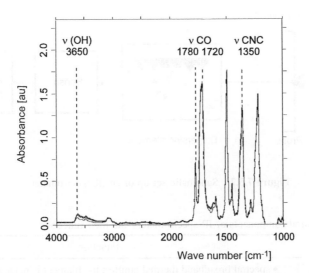

Figure 7.5.1   Infrared spectrum of polyimide PI 2540. The absorption band at $3650\,\mathrm{cm}^{-1}$ is caused by absorbed water molecules in the polyimide (reproduced from [5] with permission of *Journal of the Electrochemical Society*)

oscillating mass, group oscillations are different from skeletal oscillations regarding their wavelength range.

Figure 7.5.1 provides an example of the infrared spectrum of polyimide. We can clearly discern the characteristic absorption bands of the characteristic polar groups (=C=0- or C=N–C-groups of the imide ring). In Figure 1.1, the transmission of the atmosphere was determined in a similar way, with the absorption bands being determined by the absorption due to humidity and CO_2 – as well as N_2 content of the air.

A comparison with the spectra of known combinations allows us to gain from the infrared spectrum a lot of information regarding the structure of a chemical compound. For very large molecules (e.g. polymers), the complete interpretation of a spectrum is often difficult or even impossible. Therefore, we often only assign the strongest bands and identify some of the weaker bands as harmonic or combination oscillations.

7.5.2   Design of an Infrared Spectrometer

Infrared spectrometers are used for the qualitative and quantitative determination of the interaction of infrared radiation and matter. Main applications are infrared spectrometers for absorption and reflection experiments that intend to explain the structure of chemical species. Figure 7.5.2 presents the measuring principle. IR spectrometers usually consist of the structural components represented in Table 7.5.4.

Table 7.5.5 presents the most important types of infrared spectrometers:

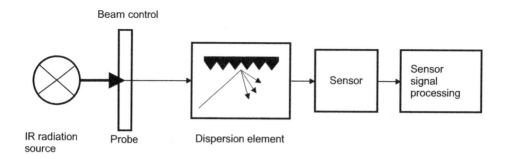

Figure 7.5.2   Schematic set-up of an IR spectrometer

Table 7.5.4   Structural components of IR spectrometers and their functions

Component	Function
IR radiation source	• spectral broadband thermal emitter; the filament is made red-hot or heated to incandescence, • typical: 'NERNST rod' made of oxides of rare earths or 'Globar' consisting of silicon carbide.
Beam control (propagation path)	• guiding and concentrating of the emitter by aluminium- or silver-vaporised mirrors (prevents the absorption occurring in lenses and common mirrors), • focal point of the emitter in the probe.
Window (propagation path)	• surrounds the probe and protects the detector, • consisting of IR-transparent mineral salts: NaCl or KBr for wavenumbers above 650 or $400\,cm^{-1}$, • for aqueous solutions: AgCl cuvettes ($>430\,cm^{-1}$) or CaF_2 cuvettes ($<1200\,cm^{-1}$).
Probe	• Fluids: confined between IR-transparent mineral salt plates (similar to the windows), layer thicknesses of about $10\,\mu m$, for diluted solutions 0.1–10 mm, in order to achieve a sufficiently large absorption length, • gases: cuvettes (glass cells) of a length of 5–10 cm, at the ends closed by alkali halide windows. At low pressures, we achieve large absorption lengths for gases using multiple reflection in the cuvette, • solid substances: mainly in solvents or as suspension to prevent reflection and scattering of the IR radiation at individual particles.
Dispersion element (propagation path)	• mainly gratings due to the smaller absorption in comparison to prisms (grating spectrometers), • revolving grating for projecting the wavelength-dependent diffracted radiation to single-element sensor, fixed gratings at the detector lines.
Sensor	• mainly pyroelectric sensors or thermopiles (compare Sections 6.4 and 6.5). Thermal sensors show only a limited wavelength/wave number dependence over the entire measuring range.
Sensor signal processing	• amplifies with the lowest noise possible (compare Examples 6.5.3 and 6.5.4 for pyroelectric sensors, amongst other), • calculates the spectrum (e.g. for FTIR spectrometers (see Table 7.5.5) applying FOURIER transform based on the time signal).

Table 7.5.5 Types of infrared spectrometers

Type	Structure

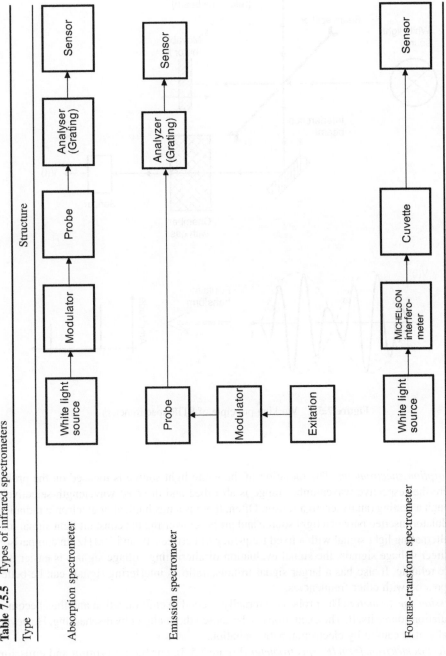

Absorption spectrometer: White light source → Modulator → Probe → Analyser (Grating) → Sensor

Emission spectrometer: Exitation → Modulator → Probe → Analyzer (Grating) → Sensor

FOURIER-transform spectrometer: White light source → MICHELSON interfero-meter → Cuvette → Sensor

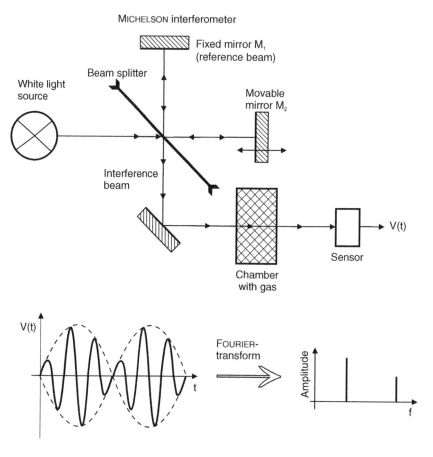

Figure 7.5.3 Working principle of FTIR spectrometers

- *Absorption spectrometer.* The radiation of the white light source is focused on the probe, where the respective wavenumber range is absorbed and directed wavelength-selectively through a grating (analyser) to a sensor. Often, there is a mechanical or electronic radiation modulator inserted between light source and probe, converting the constant light signal into an alternating light signal with a fixed frequency of between 10 and 100 Hz. In comparison to direct voltage signals, the signal evaluation of alternating voltage signals is easier and more reliable. It also has a larger signal-to-noise ratio as interfering signals can be better suppressed with other frequencies.
- *Emission spectrometer.* The probe is externally excited to emit radiation and thus becomes a radiation source itself. The excitation can be caused thermally or by discharging. However, usually, it is caused by electromagnetic radiation.
- *FTIR (FOURIER transform IR) spectrometer* (Figure 7.5.3). For the absorption and emission spectrometry, due to the diffraction of the dispersion element only IR light of a specific

wavelength reaches the sensor. In order to record the complete spectrum, we have to cover the entire spectrum which requires long measuring times. Applying a MICHELSON interferometer with a moveable mirror, causes constructive and destructive interference with the reference radiation. The moveable mirror M_2 causes time signal $V(t)$ at the sensor which now can be FOURIER-transformed and after a short measuring time provides the complete wavelength or wavenumber absorption spectrum of the probe. The FOURIER transform spectrometer does not require any dispersive elements, such as prisms or gratings. FTIR spectrometry has the following advantages [4,6]:

- high responsivity,
- short measuring times in the range of seconds for the complete spectrum,
- enables dynamic processes, such as for gas or fluid chromatography,
- larger signal-to-noise ratio due to the simultaneous absorption of all wavenumbers falling within a certain wavenumber range and to a highly reliable divergence of the radiation in the interferometer,
- results in a high spectral resolution capability.

Due to its high responsivity and good spectral resolution capability, it is possible to record interpretable spectra even for highly absorbing probes, diffuse reflectance of the probe and complicated probe arrangements (e.g. for attenuated total reflectance, ATR).

7.6 Gas Analysis

The infrared gas analysis can be considered a special operating case of infrared spectrometers where we do not record the spectrum, but determine the transmission for different wavelengths only. We select wavelengths that correspond to the characteristic absorption bands of the gas components. Figure 7.6.1 shows the principal structure of a gas analyser for several known gas components. The device usually consists of the components included in Table 7.6.1.

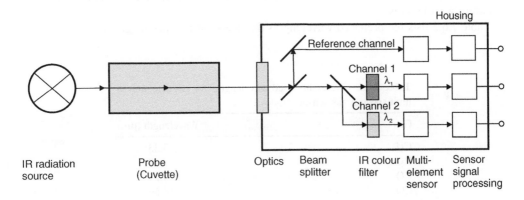

Figure 7.6.1   Schematic set-up of a multi-spectral gas sensor for two gases

Table 7.6.1 Structural components of multi-spectral gas sensors

Component	Functions
IR radiation source	• spectral broadband thermal emitter, mainly IR filament lights [7], • laser diodes with an emission wavelength that is specifically selected according to the gas to be detected (see Table 7.6.2); replaces the combination of broadband thermal emitter, beam splitter and IR colour filter (laser diode spectrometry).
Probe	• transmitting atmosphere or • gas in cuvettes (such as IR spectrometry, compare Table 7.5.4).
Optics (propagation path)	• entrance aperture for sensor, • limits the spectral range.
Beam splitter (propagation path)	• splits the radiation as evenly as possible between reference and measuring channels (mainly mirror systems, e.g. mirrored pyramids for four-channel sensors).
IR colour filter (propagation path)	• filters out the specific wavelength of the respective gas component (mainly interference filters due to its narrow-band nature and high edge steepness).
Multi-element sensor	• converts the incident (projected) radiation according to elements into an electric signal (mainly thermopiles or pyroelectric sensors, compare Section 6.4 and 6.5).
Sensor signal processing	• amplifies with the lowest noise possible (compare Examples 6.5.3 and 6.5.4 for pyroelectric sensors, amongst other), • implements the best signal-to-noise ratio possible or, respectively, maximum detectivity D (e.g. Equations 6.5.78 or 6.5.79), • compensates interferences from ambient temperature fluctuation, • compensates the effects of the atmosphere in the propagation path by reference measurements in the reference channel.

Multi-spectral gas sensors have a working principal that is similar to that of absorption spectrometers (see Table 7.5.5). As opposed to spectrometers they do not include dispersive elements, but use IR colour filters – usually interference filters – to separate wavelengths.

Table 7.6.2 lists the preferred wavelengths that are used for specific gases.

Table 7.6.2 Preferably used wavelengths for gas detection in multispectral gas sensors

Gas	Wavelength (μm)
CH_4	3.33
HC	3.40
CO_2	4.24
CO	4.66
NO_X	5.30
SO_2	7.30

References

1. Walther, L. and Gerber, D. (1981) *Infrarotmeßtechnik (Infrared Measurement Technology)*, Verlag Technik.
2. Pauli, H. and Engel, F. (1999) *Das Pyrometer-Kompendium (Pyrometer Compendium)*, IMPAC Electronic.
3. Rosch, R., Zapp, R. and Hofmann, G. (1996) *Passiv-Infrarotbewegungsmelder (Passive Infrared Motion Detector)*, Verlag Moderne Industrie.
4. Banwell, C.N. and McCash, E.M. (2008) *Fundamentals of Molecular Spectroscopy*, McGraw-Hill Higher Education.
5. Buchhold, R., Nakladal, A., Gerlach, G. *et al.* (1998) A study of the microphysical mechanisms of adsorption in polyimide layers for microelectronic applications. *Journal of the Electrochemical Society*, **145**(11), 4012–4018.
6. Herrmann, K. and Walther, L. (1990) *Wissensspeicher Infrarottechnik (Store of Knowledge in Infrared Technology)*, Fachbuchverlag.
7. Elias, B.C. (2008) Infrared emitters for spectroscopic applications. *Sensor + Test 2008 Proceedings (OPTO 2008, IRS² 2008), Nürnberg, 6–8. May, AMA Fachverband für Sensorik*, pp. 237–242.
8. Budzier, H., Krause, V., Gerald, G. and Wassiliew, D. (2005) Microbolometer-based infrared camera for the 3–5 μm spectral range. Proc. SPIE, Jena, September 12–15, Vol. 5964, pp. 244–251.
9. Foote, M.C., Kenyon, M., Krueger, T.R. *et al.* (2004) Thermopile Detector Arrays for Space Science Applications, NASA Technical Reports.
10. Hanson, C.M., Beratan, H.R. and Arbuthnot, D.L. (2008) Uncooled thermal imaging with thin-film ferroelectric detectors. Infrared Technology and Applications XXXIV; Proc. of SPIE, Vol. 6940, pp. S.694025-1–694025-12.
11. Hofmann, G., Norkus, V., Budzier, H. *et al.* (1995) Uncooled pyroelectric arrays for contactless temperature measurements. *Proceedings of SPIE*, **2474**, 98–109.
12. Koepernik, J., Budzier, H. and Hofmann, G. (1995) Influence of non-ideal chopper design on nonuniformity in unrefrigerated pyroelectric staring array systems. *Proceedings of SPIE*, **2552**, 624–635.
13. Wallrabe, A. (2001) *Nachtsichttechnik (Night Vision Technology)*, Vieweg&Sohn Verlag.
14. Schultz, M. and Caldwell, L. (1995) Nonuniformity correction and correctability of infrared focal plane arrays. *Infrared Physics and Technology*, **36**, 763–777.
15. Horny, N. (2003) FPA camera standardisation. *Infrared Physics and Technology*, **44**, 109–119.

Appendix A: Constants

Denomination	Symbol	Value
Elementary charge	e	1.602176×10^{-19} As
Permeability of vacuum	μ_0	$4\pi \cdot 1 \times 10^{-7}$ Vs/(A m) $= 12.56637 \times 10^{-7}$ Vs/(A m)
Dielectric constant of vacuum	ε_0	8.854188×10^{-12} As/(V m)
Speed of light in vacuum	c_0	2.99792×10^{8} m/s
AVOGADRO constant	L	6.022142×10^{23} mol^{-1}
BOLTZMANN constant	k_B	1.380650×10^{-23} Ws/K
STEFAN-BOLTZMANN constant	σ	5.670512×10^{-8} W/(m²K⁴)
Absolute zero temperature		$-273.15\,°C = 0\,K$
Standard atmospheric pressure	p_0	$101{,}325$ N/m²
PLANCK's constant	h	6.626069×10^{-34} Ws²
First radiation constant	c_1	3.741775×10^{-16} Wm²
Second radiation constant	c_2	1.438769×10^{-2} m K

Thermal Infrared Sensors: Theory, Optimisation and Practice   Helmut Budzier and Gerald Gerlach
© 2011 John Wiley & Sons, Ltd

Appendix B: PLANCK's Law of Radiation and Derived Laws

A cubic cavity with edge length L is assumed to have constant temperature T. The inner walls of the cavity are ideally mirror-coated. In the cavity, there are photons that can be – due to the particle structure of light – interpreted as photon gas that is in thermal equilibrium with the cavity. Mean number $\bar{n}$ of the photons that exactly have energy $h\nu$ is given by the BOSE-EINSTEIN distribution:

$$\overline{n(\nu)} = \frac{1}{e^{\frac{h\nu}{k_B T}} - 1}. \tag{B.1}$$

Energy $h\nu$ of theses photons corresponds to the light with frequency ν or, respectively, with wavelength

$$\lambda = \frac{c}{\nu}. \tag{B.2}$$

In the cavity, only stationary waves that have a node at the edge of the cavity can be form parallel to the walls. It thus applies that

$$L = m\frac{\lambda}{2} \text{ with } m = 1, 2, 3, \ldots \tag{B.3}$$

or, respectively, for the amplitude of wave vector k_m

$$k_m = \frac{2\pi}{\lambda_m} = \frac{\pi}{L}m. \tag{B.4}$$

Frequency ν of the light is linearly related to the amplitude of wave vector k of the light wave:

$$\nu = \frac{\omega}{2\pi} = \frac{ck}{2\pi}. \tag{B.5}$$

Thermal Infrared Sensors: Theory, Optimisation and Practice   Helmut Budzier and Gerald Gerlach
© 2011 John Wiley & Sons, Ltd

The number of the possible wave vectors corresponds to the number of modes N. It can be calculated in the reciprocal space, the k-space. Similar to wave processes, where frequency v corresponds to the reciprocal of time (typically expressed by harmonic function $\cos(\omega t)$), here wavenumber k is reciprocal to spatial coordinate x (for space- and time-related wave functions, it is described by $\cos(\omega t \pm kx)$). In the k-space, a "volume" of

$$V_k = k^3 = \left(\frac{\pi}{L}\right)^3. \tag{B.6}$$

corresponds to each possible k-value. The number of modes $N(k)$ is calculated as the quotient of the total volume of the k-space $\left(\frac{4}{3}\pi k^3\right)$ and volume V_k of the modes:

$$N(k) = 2 \cdot \frac{1}{8} \cdot \frac{\frac{4}{3}\pi k^3}{\left(\frac{\pi}{L}\right)^3} = \frac{V}{3\pi^2} k^3 \tag{B.7}$$

with V being the volume of the cavity ($V = L^3$). Factor 2 in Equation (B.7) takes into account that there are two polarisation directions per wave vector (forward and backward wave). Divisor 8 in Equation (B.7) states that only the positive octant in the k-space can be used because only positive wave vectors k make sense physically. Inserting Equation (B.5) into Equation (B.7), we arrive at:

$$N(\omega) = \frac{V}{3\pi^2}\frac{\omega^3}{c^3}. \tag{B.8}$$

We arrive at number dN of photons, that have an energy between $h\omega$ and $h(\omega + d\omega)$ or, respectively, a frequency between ω and $\omega + \Delta\omega$ results, by differentiating Equation (B.8):

$$dN = \frac{V}{\pi^2}\frac{\omega^2}{c^3}d\omega. \tag{B.9}$$

It applies for frequency-dependent energy density $u(\omega)$ that

$$u(\omega) = \frac{dE(\omega)}{dV} = \frac{d(\bar{E}N)}{V\,d\omega} \tag{B.10}$$

with mean energy $\bar{E}$ of the photons. With Equation (B.9), Equation (B.10) becomes

$$u(\omega) = \bar{E}\frac{\omega^2}{\pi^2 c^3} = \bar{E}\frac{4v^2}{c^3}. \tag{B.11}$$

As, in the following, we preferably use frequency v as a variable, it follows for Equation (B.11):

$$u(v) = \bar{E}\frac{8\pi v^2}{c^3}. \tag{B.12}$$

The relation in Equation (B.12) also derives from electrodynamics. Assuming that each photon only swings in one direction it constitutes a linear oscillator. The inner energy of a linear oscillator now is equal to its oscillation energy $\bar{E}$. Equation (B.12) reflects the interrelation between mean energy $\bar{E}$ of a linear oscillator generating electromagnetic waves in the frequency interval v to $v + dv$ and energy density $u(v)$ of this radiation. $u(v)dv$ corresponds to the number of the natural oscillations in the frequency interval v to $v + dv$.

Inserting $\bar{E} = k_B T$ for the mean energy of a photon, Equation (B.11) results in the RAYLEIGH-JEANS formula:

$$u(\omega) = \frac{k_B T}{\pi^2 c^3} \omega^2. \tag{B.13}$$

This is not possible as there would result an indefinitely high energy as we calculate total energy by integrating Equation (B.13) over the frequency of $\omega = 0$ to $\omega = \infty$. This situation is known as the ultraviolet catastrophe. PLANCK solved this problem by assuming that the photons can only adopt discrete energies, which means:

$$E_m = m h v. \tag{B.14}$$

We can then arrive at the mean energy using the BOSE-EINSTEIN distribution in Equation (B.1):

$$\bar{E} = \frac{hv}{e^{\frac{hv}{k_B T}} - 1}. \tag{B.15}$$

For the spectral energy density, with Equation (B.12) we arrive at the famous PLANCK formula:

$$u(v) = 8\pi \frac{hv^3}{c^3} \frac{1}{e^{\frac{hv}{k_B T}} - 1}. \tag{B.16}$$

It describes the energy that an ideal emitter with temperature T emits in the frequency interval between v and $v + dv$ and per volume V.

Infrared measuring technology usually describes the radiation emitted by a body using spectral exitance M_v which states the radiation power of an area element in half-space:

$$M_v = \frac{d\Phi_v}{dA}. \tag{B.17}$$

At first, Equation (B.16) is transformed in order to calculate an area-related energy:

$$E_v = u(v)dV = u(v)dA\, dx. \tag{B.18}$$

The spectral radiation flux is calculated from the energy

$$\Phi_v = \frac{dE_v}{dt} = \frac{u(v)dA\, dx}{dt}. \tag{B.19}$$

As radiation always propagates at the speed c of light, it applies that:

$$dx = cdt. \tag{B.20}$$

Now, the radiant flux per area element becomes

$$M_{v\uparrow} = \frac{d\Phi_v}{dA} = cu(v). \tag{B.21}$$

Equation (B.21) describes – applying the assumption in Equation (B.20) – radiant flux $M_{\nu\uparrow}$ that is perpendicularly emitted from the area. What we are looking for, however, is the radiant flux into the half-space. For the radiation flux that is emitted at angle α in relation to the area normal and azimuth angle γ it applies that

$$M_\nu(\alpha, \gamma) = cu(\nu)\cos \alpha. \tag{B.22}$$

The spectral exitance now is the mean value over the half-space:

$$M_\nu = \frac{1}{4\pi} \int_0^{2\pi} \int_0^{\frac{\pi}{2}} cu(\nu)\cos \alpha \sin \alpha \, d\alpha \, d\gamma. \tag{B.23}$$

As $u(\nu)$ is not related to angles α and γ, the integrals can be solved independently of each other:

$$\int_0^{\frac{\pi}{2}} \cos \alpha \sin \alpha \, d\alpha = \frac{1}{2} \tag{B.24}$$

and

$$\int_0^{2\pi} d\gamma = 2\pi. \tag{B.25}$$

Equation (B.22) then results in

$$M_\nu = \frac{c}{4}u(\nu). \tag{B.26}$$

Inserting Equation (B.16) we arrive at PLANCK's law:

$$M_\nu = 2\pi \frac{h\nu^3}{c^2} \frac{1}{e^{\frac{h\nu}{k_B T}} - 1}. \tag{B.27}$$

PLANCK's radiation law is often stated as a function of wavelength. It follows for the conversion of M_ν into M_λ that

$$M_\nu d\nu = M_\lambda d\lambda = \frac{c}{4}u(\nu)d\nu \tag{B.28}$$

or, respectively,

$$M_\lambda = \frac{c}{4}u(\nu)\frac{d\nu}{d\lambda} \tag{B.29}$$

with

$$\nu = \frac{c}{\lambda}. \tag{B.30}$$

It results that

$$\left|\frac{dv}{d\lambda}\right| = \frac{c}{\lambda^2}.$$ (B.31)

Thus, for spectral exitance we arrive at the most familiar form of PLANCK's law:

$$M_\lambda = 2\pi \frac{hc^2}{\lambda^5} \frac{1}{e^{\frac{hc}{\lambda k_B T}} - 1}$$ (B.32)

We usually summarize the constants in Equation (B.32):

$$M_\lambda = \frac{c_1}{\lambda^5} \frac{1}{e^{\frac{c_2}{\lambda T}} - 1}$$ (B.33)

with the so-called radiation constants

$$c_1 = 2\pi hc^2 = 3,741832\text{E-}16 \text{ W} \cdot \text{m}^2$$ (B.34)

and

$$c_2 = \frac{hc}{k_B} = 1,438786\text{E-}2 \text{ K} \cdot \text{m}.$$ (B.35)

For total exitance M_{BB}, PLANCK's law has to be integrated over all frequencies:

$$M_{BB} = \int_0^\infty M_v \, dv.$$ (B.36)

Substituting

$$x = \frac{h}{k_B T} v$$ (B.37)

and with

$$dv = \frac{k_B T}{h} dx$$ (B.38)

we arrive at

$$M_{BB} = \frac{2\pi (k_B T)^4}{c^2 h^3} \int_0^\infty \frac{x^3}{e^x - 1} dx$$ (B.39)

There is a known solution for the integral:

$$\int_0^\infty \frac{x^3}{e^x - 1} dx = \frac{\pi^4}{15}.$$ (B.40)

Thus, Equation (B.39) becomes the STEFAN-BOLTZMANN law:

$$M_{BB} = \frac{2\pi^5 k_B^4}{15c^2 h^3} T^4 = \sigma T^4 \tag{B.41}$$

with the STEFAN-BOLTZMANN constant

$$\sigma = \frac{2\pi^5 k_B^4}{15c^2 h^3} = 5,67\text{E}{-}8\,\frac{\text{W}}{\text{m}^2\text{K}^4}. \tag{B.42}$$

We have to set the derivative of PLANCK's law (Equation (B.27)) to zero in order to calculate the radiation maximum:

$$\frac{dM_v}{dv} = 0 \tag{B.43}$$

Therefore, we use Equation (B.37) for a substitution and arrive at

$$\left(2\pi \frac{(k_B T)^3}{h^2}\frac{x^3}{e^x-1}\right)' = 0 \tag{B.44}$$

The derivative results in

$$\frac{3x^2}{e^x-1} - \frac{x^3 e^x}{(e^x-1)^2} = 0. \tag{B.45}$$

From that, it follows

$$(3-x)e^x = 3. \tag{B.46}$$

The LAMBERT-W function W_L is suitable to solve this equation[1]. For this purpose, we have to rearrange Equation (B.46). Extending by $-e^{-3}$, Equation (B.46) becomes

$$(x-3)e^{x-3} = -3e^{-3} \tag{B.47}$$

The solution of this Equation is

$$x = 3 + W_L(-3e^{-3}) = 2,821439. \tag{B.48}$$

With Equation (B.37), we arrive at WIEN's displacement law:

$$v_{\text{max}}(T) = 2,821\frac{k_B}{h}T = 5,878\text{E}10\,\frac{\text{Hz}}{\text{K}}T. \tag{B.49}$$

---

[1] The LAMBERT-W function solves Equation $w(x)e^{w(x)} = z$ in the form $w(x) = W_L(z)$. It is also called Omega function as $W_L(1) = $ Omega constant.

Maximum wavelength λ_{max} - the wavelength where spectral exitance M_λ has a maximum –
cannot be calculated by simply using Equation (B.30) to rearrange Equation (B.49) for
wavelength as wavelength λ and frequency v are not linearly related. This means that the
intervals $\lambda + d\lambda$ and $v + dv$ are not equal and that there is no common maximum. In order
to calculate maximum wavelength λ_{max}, the derivative of Equation (B.32) has to be set to zero.
With the abbreviation

$$A = \frac{hc}{k_B T} \tag{B.50}$$

we have to calculate the following term:

$$\left(2\pi c k_B T \frac{A}{\lambda^5} \frac{1}{e^{\frac{A}{\lambda}}-1} \right)' = 0. \tag{B.51}$$

After factorising all constants, Equation (B.51) becomes:

$$\left(\frac{1}{\lambda^5} \frac{1}{e^{\frac{A}{\lambda}}-1} \right)' = 0. \tag{B.52}$$

With product rule $(uv)' = u'v + uv'$ we can calculate the derivative:

$$(u)' = \left(\frac{1}{\lambda^5} \right)' = -\frac{5}{\lambda^6}, \tag{B.53}$$

$$(v)' = \left(\frac{1}{e^{\frac{A}{\lambda}}-1} \right)' = \frac{A}{\lambda^2} \frac{e^{\frac{A}{\lambda}}}{\left(e^{\frac{A}{\lambda}}-1\right)^2}, \tag{B.54}$$

$$(uv)' = -\frac{5}{\lambda^6} \frac{1}{e^{\frac{A}{\lambda}}-1} + \frac{A}{\lambda^7} \frac{e^{\frac{A}{\lambda}}}{\left(e^{\frac{A}{\lambda}}-1\right)^2} = 0. \tag{B.55}$$

Now we again insert A and simplify:

$$\frac{hc}{\lambda k_B T} \frac{1}{1-e^{-\frac{hc}{\lambda k_B T}}} - 5 = 0. \tag{B.56}$$

Substituting

$$x = \frac{hc}{\lambda k_B T}, \tag{B.57}$$

we arrive at

$$\frac{x}{1-e^{-x}} = 5. \tag{B.58}$$

By further rearranging, we arrive at

$$(x-5)e^x = -5. \tag{B.59}$$

Expanding by e^{-5}, it applies that:

$$(x-5)e^{x-5} = -5e^{-5}. \tag{B.60}$$

The solution of this Equation is:

$$x = 5 + W_L(-5e^{-5}) = 4,965114. \tag{B.61}$$

With Equation (B.57), we arrive at WIEN's displacement law:

$$\lambda_{max}(T) = \frac{hc}{4,965k_BT} = \frac{2,898 \times 10^{-3}\,\text{K m}}{T}. \tag{B.62}$$

The product of temperature and the wavelength of the maximum exitance is constant:

$$\lambda_{max}T = 2898\,\text{K μm}. \tag{B.63}$$

If we now made the mistake to calculate λ_{max} directly from Equation (B.49), with Equation (B.30) we would arrive at the completely wrong value of 5.1×10^{-3} K·m/T.

Appendix C: Calculation of the Solid Angle of a Rectangular Area

Rectangular area A is assumed to be situated in the xy-plane. Point P is situated at any point in relation to that (Figure C.1).

We need area $d\vec{A}$ in order to calculate its solid angle in a parameterized representation:

$$d\vec{A} = \frac{\delta\vec{r}}{\delta u} \times \frac{\delta\vec{r}}{\delta v}. \tag{C.1}$$

The solid angle generally results in

$$\Omega = \iint\limits_{vu} \frac{\vec{R}}{|\vec{R}|^3} \left(\frac{\delta\vec{r}}{\delta u} \times \frac{\delta\vec{r}}{\delta v} \right) du \, dv. \tag{C.2}$$

The following relations apply:

$$\vec{R} = \vec{r} - \vec{r}_0, \tag{C.3}$$

$$\vec{r}_0 = \begin{pmatrix} x_0 \\ y_0 \\ z_0 \end{pmatrix}, \tag{C.4}$$

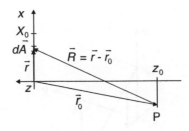

Figure C.1 Arrangement for calculating the solid angle of a rectangular area

Thermal Infrared Sensors: Theory, Optimisation and Practice Helmut Budzier and Gerald Gerlach
© 2011 John Wiley & Sons, Ltd

$$\vec{r} = \begin{pmatrix} u \\ v \\ 0 \end{pmatrix} \quad \begin{matrix} U_1 \le u \le U_2 \\ V_1 \le v \le V_2 \end{matrix}, \tag{C.5}$$

$$\vec{R} = \begin{pmatrix} u-x_0 \\ v-y_0 \\ z_0 \end{pmatrix}, \tag{C.6}$$

$$|\vec{R}| = \sqrt{x_0^2 + y_0^2 + z_0^2 + v^2 + u^2 - 2(ux_0 + vy_0)}, \tag{C.7}$$

$$\frac{\delta \vec{r}}{\delta v} = \begin{pmatrix} 0 \\ 1 \\ 0 \end{pmatrix}, \tag{C.8}$$

$$\frac{\delta \vec{r}}{\delta u} = \begin{pmatrix} 1 \\ 0 \\ 0 \end{pmatrix}, \tag{C.9}$$

$$d\vec{A} = \frac{\delta \vec{r}}{\delta u} \times \frac{\delta \vec{r}}{\delta v} = \begin{vmatrix} \vec{e}_x & \vec{e}_y & \vec{e}_z \\ 1 & 0 & 0 \\ 0 & 1 & 0 \end{vmatrix} = \vec{e}_z, \tag{C.10}$$

$$\vec{R}d\vec{A} = z_0, \tag{C.11}$$

$$\Omega_{RE} = \iint\limits_{uv} \frac{z_0}{\left(x_0^2 + y_0^2 + z_0^2 + v^2 + u^2 - 2ux_0 - 2vy_0\right)^{\frac{3}{2}}} du\, dv. \tag{C.12}$$

In order to arrive at manageable and solvable terms, we locate spatial point P at a corner of the rectangle on the z-axis (Figure C.2):

$$x_0 = y_0 = 0.$$

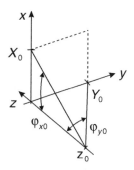

Figure C.2   Arrangement of Figure C.1 with spatial point P on the z-axis

Thus, Equation (C.12) becomes

$$\Omega_{RE} = z_0 \iint\limits_{uv} \frac{1}{\left(z_0^2 + v^2 + u^2\right)^{\frac{3}{2}}} \, du \, dv. \tag{C.13}$$

It results from a first integration that

$$\int_0^{X_0} \frac{du}{\left(z_0^2 + v^2 + u^2\right)^{\frac{3}{2}}} = \frac{u}{\left(z_0^2 + v^2\right)\sqrt{z_0^2 + v^2 + u^2}}\Bigg|_0^{X_0} = \frac{X_0}{\left(z_0^2 + v^2\right)\sqrt{z_0^2 + v^2 + X_0^2}} \tag{C.14}$$

and the solid angle integral becomes

$$\Omega_{RE} = z_0 X_0 \int_0^{Y_0} \frac{dv}{\left(z_0^2 + v^2\right)\sqrt{z_0^2 + v^2 + X_0^2}}. \tag{C.15}$$

Substituting v

$$v = z_0 \tan \varphi \text{ with } 0 \leq \varphi \leq \varphi_{Y0} \tag{C.16}$$

and with the relations

$$\tan \varphi_{Y0} = \tan \frac{Y_0}{z_0}, \tag{C.17}$$

$$\sin \varphi_{Y0} = \frac{Y_0}{A_Y} \tag{C.18}$$

we arrive at

$$dv = \frac{z_0}{\cos^2 \varphi} d\varphi, \tag{C.19}$$

$$\frac{1}{\cos^2 \varphi} = 1 + \tan^2 \varphi, \tag{C.20}$$

$$A_X^2 = z_0^2 + X_0^2, \tag{C.21}$$

$$A_Y^2 = z_0^2 + Y_0^2. \tag{C.22}$$

With Equation (C.17) to (C.22), it results for the integral in Equation (C.15) that

$$
\int_0^{Y_0} \frac{dv}{(z_0^2 + v^2)\sqrt{z_0^2 + v^2 + X_0^2}} = \int_0^{\varphi_{Y0}} \frac{z_0}{\cos^2\varphi \left(z_0^2(1+\tan^2\varphi)\right)\sqrt{z_0^2(1+\tan^2\varphi) + X_0^2}} d\varphi
$$

$$
= \int_0^{\varphi_{Y0}} \frac{\cos\varphi}{z_0 X_0 \sqrt{\dfrac{z_0^2}{X_0^2} + \cos^2\varphi}} d\varphi = \int_0^{\varphi_{Y0}} \frac{\cos\varphi}{z_0 X_0 \sqrt{\dfrac{z_0^2}{X_0^2} + 1 - \sin^2\varphi}} d\varphi = \frac{1}{z_0 X_0} \arcsin\frac{\sin\varphi}{\sqrt{\dfrac{z_0^2}{X_0^2} + 1}}\Bigg|_0^{\varphi_{Y0}}
$$

$$
= \frac{1}{z_0 X_0} \arcsin\frac{\sin\varphi_{Y0}}{\sqrt{\dfrac{z_0^2 + X_0^2}{X_0^2}}}. \tag{C.23}
$$

Now the solid angle in Equation (C.15) becomes

$$
\Omega_{RE} = \arcsin\sqrt{\frac{X_0^2 Y_0^2}{(z_0^2 + X_0^2)(z_0^2 + Y_0^2)}} = \arcsin\frac{X_0 Y_0}{A_X A_Y}. \tag{C.24}
$$

Rearranging this Equation, we also arrive at

$$
\Omega_{RE} = \arcsin\frac{1}{\sqrt{\left(\dfrac{z_0^2}{X_0^2}+1\right)\left(\dfrac{z_0^2}{Y_0^2}+1\right)}} = \arcsin\left[\sin(\varphi_{x0})\sin(\varphi_{y0})\right]. \tag{C.25}
$$

With $X_0 = Y_0 \to \infty$, the argument becomes 1 and thus Ω_{RE} reaches its maximum:

$$
\Omega_{RE,\max} = \frac{\pi}{2}\Omega_0. \tag{C.26}
$$

This case comprises one fourth of the half-space which corresponds to the above definition of the calculated area:

$$
\Omega_{HR} = 4*\Omega_{RE,\max} = 2\pi\Omega_0. \tag{C.27}
$$

Further Reading and Sources

Further Reading

Acceta, J.S. and Shumaker, D.L. (eds) (1993) *The Infrared and Electro-Optical Systems Handbook*, SPIE Optical Engineering Press, Bellingham.
Volume 1: G.J. Zisses: Sources of Radiation
Volume 2: F.G. Smith: Atmospheric Propagation of Radiation
Volume 3: W.D. Rogatto: Electro-Optical Components
Volume 4: M.C. Dudzik: Electro-Optical Systems Design, Analysis and Testing
Volume 5: S.B. Campana: Passive Electro-Optical Systems
Volume 6: C.S. Fox: Active Electro-Optical Systems
Volume 7: D.H. Pollock: Countermeasure Systems
Volume 8: S.R. Robinson: Emerging Systems and Technologies.
Akin, T. (2005) *CMOS-based Thermal Sensors: Advanced Micro and Nanosystems*, vol. **2**, CMOS-MEMS, Wiley-VCH Verlag, Weinheim.
Capper, P. and Elliott, C.T. (2001) *Infrared Detectors and Emitters: Materials and Devices*, Kluwer Academic Publisher, Dordrecht.
Daniels, A. (2007) *Field Guide to Infrared Systems*, SPIE Press, Bellingham.
Holst, G.C. (1998) *Testing and Evaluation of Infrared Imaging Systems*, JCD Publishing Company, Bellingham.
Infrared Physics and Technology, Elsevier, London.
Kimata, M. (1998) *Infrared Focal Plane Arrays*, Sensors Update, vol. **4** (eds H. Baltes, W. Göpel and J. Hesse) Wiley-VCH, Weinheim.
Lang F. S.B. and Das-Gupta, D.K. (2001) Pyroelectricity: fundamentals and applications, in *Handbook of Advanced Electronic and Photonic Materials and Devices*, vol. **4** (ed. H.S. Nalva) Academic Press, London.
Müller, R. (1989) *Rauschen (Noise)*, Springer-Verlag, Berlin.
Rogalski, A. (2003) Infrared detectors: status and trends. *Progress in Quantum Electronics*, **27**, 59–210.
Rohsenow, W.M., Hartnett, J.P. and Cho, Y.I. (1998) *Handbook of Heat Transfer*, McGraw-Hill, New York.
Schuster, N. and Kolobrodov, V.G. (2004) *Infrarotthermografie (Infrared Thermography)*, 2nd edn, Wiley-VCH Verlag, Weinheim.
Stahl, K. and Miosga, G. (1986) *Infrarottechnik (Infrared Technology)*, Hüthig Verlag, Heidelberg.
Wolfe, W.L. (1996) *Introduction to Infrared Design*, SPIE Press, Bellingham.

Wersing, W. and Bruchhaus, R. (2000) Pyroelectric devices and applications, in D.J. Deborah (2000) *Handbook of Thin Film Devices*, Vol. **5** Academic Press, San Diego.
Williams, T.L. (2009) *Thermal Imaging Cameras*, CRC Press, Boca Raton.
Vollmer, M. and Möllmann, K.-P. (2010) Infrared Thermal Imaging, Wiley-VCH, Weinheim.

Congresses and Conferences on Infrared Sensors

IRS[2] Conference, Nürnberg: http://www.sensor-test.de/.
Freiburger Infrarotkolloquium: http://www.infrared-workshop.de/.
Infrared Technology and Applications: http://spie.org.

Congresses and Conferences on Infrared Sensor Applications

ThermoSense, Florida: http://www.thermosense.org.
QIRT: Quantitative InfraRed Thermography: http://qirt.gel.ulaval.ca.

Index

acceleration 194, 206
AIRY disc 126
autocorrelation function 79, 83

bimorphous sensor 236
blackbody 33, 53, 97
BLIP 114, 116, 119
bolometer 150, 175, 196, 219

capacity, heat 155
CARSON theorem 85, 93
charge density 16, 197
coefficient, piezoelectric 16, 193, 207
  pyroelectric 16, 192
conductance, thermal 155
conductivity 165, 177
cross correlation 80
current density 16, 177

dead pixels 112
dielectric displacement 16, 192

electric field strength 16, 192
energy, radiation 27
equivalent noise bandwidth 84
ergodicity 69
exitance, spectral 27, 34

ferroelectrics 191
field of view 122
flux, magnetic 16
  radiant 27
f-number 56, 123
form factor 53

FRAUNHOFER diffraction 125
FRESNEL lens 273

GAUSSIAN distribution 75
GOLAY cell 242

IETD 269
IFOV 122
infrared camera 260
  spectrometry 274
  far 4
  mid 4
  near 4
intensity 17, 28
  spectral 29
irradiance 28, 54
irradiation 29

LAMBERTIAN radiators 51

magnetic field strength 16
MAXWELL's equations 15
mechanical stress 16, 192, 207
microbolometer 175, 219
microphony 206
modulation transfer function 127
MTF, capacitive 138
  geometrical 133, 168
  optical 131
  thermal 135, 168, 171

NEP 112
NETD 117

Thermal Infrared Sensors: Theory, Optimisation and Practice Helmut Budzier and Gerald Gerlach
© 2011 John Wiley & Sons, Ltd

Printed and bound by CPI Group (UK) Ltd, Croydon, CR0 4YY